Organic Fertilizers: Types, Production and Impacts on Environment

Organic Fertilizers: Types, Production and Impacts on Environment

Dr. Arvind Kumar

RANDOM PUBLICATIONS
NEW DELHI (INDIA)

Organic Fertilizers: Types, Production and Impacts on Environment

ISBN 978-93-5111-599-1

Published in 2015 in India by

RANDOM PUBLICATIONS

4376-A/4B, Gali Murari Lal, Ansari Road
New Delhi-110 002
Phone : +9111-43580356, 011-23289044, 011-43142548
e-mail: sales@randompublications.com,
info@randompublications.com, randomexports@gmail.com

Reprinted 2021

Type Setting by : Friends Media, Delhi-110089
Digitally Printed at : Replika Press Pvt. Ltd.

Preface

Organic farming is an agricultural philosophy and a farm management system aiming at promotion of plant, animal and human health. Organic farming is a farming integration of biological, cultural and natural inputs, including integrated diseases and pest management practices.

Organic fertilizers are fertilizer compounds that contain one or more kinds of organic matter. The ingredients may be animal or vegetable matter or a combination of the two. It is possible to purchase commercial brands of organic rich fertilizer as well as prepare organic fertilizer at home by building a compost heap.

When describing fertilizer, it is important to remember that the working definition of organic is the same as that applied to organic foods. That is, the fertilizer is composed of elements that are produced in a completely natural manner, without the aid of any synthetically manufactured components or additives.

Many organic materials serve as both fertilizers and soil conditioners—they feed both soils and plants. This is one of the most important differences between a chemical approach and an organic approach toward soil care and fertilizing. Soluble chemical fertilizers contain mineral salts that plant roots can absorb quickly.

However, these salts do not provide a food source for soil microorganisms and earthworms, and will even repel earthworms because they acidify the soil. Over time, soils treated only with synthetic chemical fertilizers lose organic matter and the all-important living organisms that help to build a quality soil.

Biofertilizer is defined as a substance which contains living organisms which, when applied to seed, plant surface, or soil, colonize the rhizosphere or interior of the plant and promote growth by increasing the supply or availability of primary nutrients to the host plant. Biofertilizers are well recognized as an important component of integrated plant nutrient management for sustainable agriculture and hold a great promise improve crop yield.

Aiming at the increasing practice of organic farming and use of bio-fertilizers, the book is of tremendous use to students, teachers, scholars, farmers and general readers.

I would like to thank my team for standing beside me throughout my career and writing this book. My special thanks go to "Random Publications" who have published the book.

–Dr. Arvind Kumar

Contents

1

Introduction to Fertilizers

FERTILIZER

Fertilizer (or fertiliser) is any material of natural or synthetic origin (other than liming materials) that is applied to soils or to plant tissues (usually leaves) to supply one or more plant nutrients essential to the growth of plants. Conservative estimates report 30 to 50% of crop yields are attributed to natural or synthetic commercial fertilizer. Global market value is likely to rise to more than US$185 billion until 2019. The European fertilizer market will grow to earn revenues of approx. €15.3 billion in 2018.

MECHANISM

Fertilizers enhance the growth of plants. This goal is met in two broad ways, the traditional one being additives that provide nutrients. The second mode by which some fertilizers act is to enhance the effectiveness of the soil by modifying its water retention and aeration. This article, like most on fertilizers, emphasizes the nutritional aspect. Fertilizers typically provide, in varying proportions:

- Three main macronutrients: nitrogen (N), phosphorus (P), potassium (K);
- Three secondary macronutrients: calcium (Ca), magnesium (Mg), and sulfur (S);
- Micronutrients: copper (Cu), iron (Fe), manganese (Mn), molybdenum (Mo), zinc (Zn) and nickel (Ni), and sometimes boron (B), silicon (Si), cobalt (Co), vanadium (V).

The nutrients required for healthy plant life are classified according to the elements, but the elements are not used as fertilizers. Instead compounds containing these elements are the basis of fertilizers. The macronutrients are consumed in larger quantities and are present in plant tissue in quantities from 0.15% to 6.0% on a dry matter (DM) (0% moisture) basis. Plants are made up of four main elements: hydrogen, oxygen, carbon, and nitrogen. Carbon, hydrogen and oxygen are widely available as water and carbon dioxide. Although nitrogen makes up most of the atmosphere, it is in a form that is unavailable to plants. Nitrogen is the most important fertilizer since nitrogen is present inproteins, DNA and other components (e.g., chlorophyll). To be nutritious to plants, nitrogen must be made available in a "fixed" form. Only some bacteria and their host plants (notably legumes) can fix atmospheric nitrogen (N_2) by converting it to ammonia. Phosphate is required for the production of DNA and ATP, the main energy carrier in cells, as well as certain lipids.

Micronutrients are consumed in smaller quantities and are present in plant tissue on the order of parts-per-million (ppm), ranging from 0.15 to 400 ppm DM, or less than 0.04% DM. These elements are often present at the active sites of enzymes that carry out the plant's metabolism. Because these elements enable catalysts (enzymes) their impact far exceeds their weight percentage.

CLASSIFICATION

Fertilizers are classified in many ways. They are classified according to whether they provide a single nutrient (say, N, P, or K), in which case they are classified as "straight fertilizers." "Multinutrient fertilizers" (or "complex fertilizers") provide two or more nutrients, for example N and P. Fertilizers are also sometimes classified as inorganic (the topic of most of this article) vs organic. Inorganic fertilizers excludes carbon-containing materials except ureas. Organic fertilizers are usually (recycled) plant- or animal-derived matter. Inorganic are sometimes called synthetic fertilizers since various chemical treatments are required for their manufacture.

Single nutrient ("straight") fertilizers

The main nitrogen-based straight fertilizer is ammonia or its solutions. Ammonium nitrate (NH_4NO_3) is also widely used. About 15M tons were produced in 1981, i.e., several kilograms per person. Urea is another popular source of nitrogen, having the advantage that it is a solid and non-explosive, unlike ammonia and ammonium nitrate, respectively. A few percent of the nitrogen fertilizer market (4% in 2007) is met by calcium ammonium nitrate ($Ca(NO_3)_2 \cdot NH_4NO_3 \cdot 10H_2O$).

The main straight phosphate fertilizers are the superphosphates. "Single superphosphate" (SSP) consists of 14–18% P_2O_5, again in the form of $Ca(H_2PO_4)_2$, but alsophosphogypsum ($CaSO_4 \cdot 2\ H_2O$). Triple superphosphate (TSP) typically consists of 44-48% of P_2O_5 and no gypsum. A mixture of single superphosphate and triple superphosphate is called double superphosphate. More than 90% of a typical superphosphate fertilizer is water-soluble.

Multinutrient fertilizers

These fertilizers are the most common. They consist of more than two or more nutrient components.

Binary (NP, NK, PK) fertilizers

Major two-component fertilizers provide both nitrogen and phosphorus to the plants. These are called NP fertilizers. The main NP fertilizer are monoammonium phosphate(MAP) and diammonium phosphate (DAP). The active ingredient in MAP is $NH_4H_2PO_4$. The active ingredient in DAP is $(NH_4)_2HPO_4$. About 85% of MAP and DAP fertilizers are soluble in water.

NPK fertilizers

NPK fertilizers are three-component fertilizers providing nitrogen, phosphorus, and potassium.

NPK rating is a rating system describing the amount of nitrogen, phosphorus, and potassium in a fertilizer. NPK ratings consist of three numbers separated by dashes (e.g., 10-10-10 or 16-4-8) describing the chemical content of fertilizers. The first number represents the percentage of nitrogen in the product; the second number, P_2O_5; the third, K_2O. Fertilizers do not actually contain P_2O_5 or K_2O, but the system is a conventional shorthand for the amount of the phosphorus (P) or potassium (K) in a fertilizer. A 50-pound bag of fertilizer labeled 16-4-8 contains 8 pounds of nitrogen (16% of the 50 pounds) an amount of phosphorus and potassium equivalent to that in 2 pounds of P_2O_5 (4% of 50 pounds) and 4 pounds of K_2O (8% of 50 pounds). Most fertilizers are labeled according to this N-P-K convention, though Australian convention, following an N-P-K-S system, adds a fourth number for sulfur.

Micronutrients

The main micronutrients include sources of iron, manganese, molybdenum, zinc, and copper. As for the macronutrients, these elements are provided as water-soluble salts. Iron presents special problems because it converts to insoluble (bio-unavailable) compounds at moderate soil pH and phosphate concentrations. For this reason, iron is often administered as a chelate complex, e.g. the EDTA derivative. The micronutrient needs depend on the plant. For example, sugar beets appear to require boron, and legumesrequire cobalt.

PRODUCTION

Nitrogen fertilizers

All nitrogen fertilizers are made from ammonia (NH_3), which is produced by the Haber-Bosch process. In this energy-intensive process, natural gas (CH_4) supplies the hydrogen and the nitrogen (N_2) is derived from the air. This ammonia is used as a feedstock for all other nitrogen fertilizers, such as anhydrous ammonium nitrate (NH_4NO_3) andurea ($CO(NH_2)_2$). Deposits of sodium nitrate ($NaNO_3$) (Chilean saltpeter) are also found in the Atacama desert in Chile and was one of the original (1830) nitrogen-rich fertilizers used. It is still mined for fertilizer.

Phosphates

All phosphates are obtained by extraction from minerals containing the anion PO_4^{3-}. In rare cases, fields are treated with the crushed mineral, but most often more soluble salts are produced by chemical treatment of phosphate minerals. The most popular phosphate-containing minerals are referred to collectively as phosphate rock. The main minerals are fluorapatite $Ca_5(PO_4)_3F$ (CFA) and hydroxyapatite $Ca_5(PO_4)_3OH$. These minerals are converted to water-soluble phosphate salts by treatment with sulfuric or phosphoric acids.

In the nitrophosphate process or Odda process, phosphate rock with up to a 20% phosphorus (P) content is dissolved with nitric acid (HNO_3) to produce a mixture of phosphoric acid (H_3PO_4) and calcium nitrate ($Ca(NO_3)_2$). This mixture can be combined with a potassium fertilizer to produce a *compound fertilizer* with the three macronutrients N, P and K in easily dissolved form.

Potassium

Potash is a mixture of potassium minerals used to make potassium (chemical symbol: K) fertilizers. Potash is soluble in water, so the main effort in producing this nutrient from the ore involves some purification steps, e.g. to remove sodium chloride (NaCl), common salt. Sometimes potash is referred to as K_2O, as a matter of convenience to those describing the potassium content. In fact potash fertilizers are usually potassium chloride, potassium sulfate, potassium carbonate, or potassium nitrate.

Compound fertilizers

Compound fertilizers, which contain N, P, and K, can often be produced by mixing straight fertilizers. In some cases, chemical reactions occur between the two or more components. For example monoammonium and diammonium phosphates, which provide plants with both N and P, are produced by neutralizing phosphoric acid (from phosphate rock) and ammonia (from a Haber facility):

$NH_3 + H_3PO_4 ! (NH_4)H_2PO_4$

$2\ NH_3 + H_3PO_4 ! (NH_4)_2HPO_4$

Organic fertilizer

Fig. Compost bin for small-scale production of organic fertilizer

Fig. A large commercial compost operation

The main "organic fertilizers" are, in ranked order, peat, animal wastes, plant wastes from agriculture, and sewage sludge. In terms of volume, peat is the most widely used organic fertilizer. This immature form of coal confers no nutritional value to the plants, but improves the soil by aeration and absorbing water. Animal sources include the products of the slaughter of animals. Bloodmeal, bone meal, hides, hoofs, and horns are typical components. Organic fertilizer usually contain less nutrients, but offer other advantages as well as appealing to environmentally friendly users.

Other elements: calcium, magnesium, and sulfur

Calcium is supplied as superphosphate or calcium ammonium nitrate solutions.

Top users of nitrogen-based fertilizer		
Country	**Total N use (Mt pa)**	**Amt. used for feed/pasture (Mt pa)**
China	18.7	3.0
U.S.	9.1	4.7
France	2.5	1.3
Germany	2.0	1.2
Brazil	1.7	0.7
Canada	1.6	0.9
Turkey	1.5	0.3
UK	1.3	0.9
Mexico	1.3	0.3
Spain	1.2	0.5
Argentina	0.4	0.1

APPLICATION

Fertilizers are commonly used for growing all crops, with application rates depending on the soil fertility, usually as measured by a soil test and according to the particular crop. Legumes, for example, fix nitrogen from the atmosphere and generally do not require nitrogen fertilizer.

Liquid vs solid

Fertilizers are applied to crops both as solids and as liquid. About 90% of fertilizers are applied as solids. Solid fertilizer is typically granulated or powdered. Often solids are available as prills, a solid globule. Liquid fertilizers comprise anhydrous ammonia, aqueous solutions of ammonia, aqueous solutions of ammonium nitrate and or urea. These concentrated products may be diluted with water to form a concentrated liquid fertilizer (e.g. UAN). Advantages of liquid fertilizer are its more rapid effect and easier coverage.The addition of fertilizer to irrigation water is called "fertigation".

Slow- and controlled-release fertilizers

Slow- and controlled-release involve only 0.15% (562,000 tons) of the fertilizer market. Their utility stems from the fact that fertilizers are subject to antagonistic processes. In addition to their providing the nutrition to plants, excess fertilizers can be poisonous to the same plant. Competitive with the uptake by plants is the degradation or loss of the fertilizer. Microbes degrade many fertilizers, e.g. by immobilization or oxidation. Furthermore fertilizers are lost by evaporation or leaching. Most slow-release fertilizers are derivatives of urea, a straight fertilizer providing nitrogen. Isobutylidenediurea ("IBDU") and urea-formaldehyde slowly convert in the soil to free urea, which is rapidly uptaken by plants. IBDU is a single compound with the formula $(CH_3)_2CHCH(NHC(O)NH_2)_2$ whereas the urea-formaldehydes consist of mixtures of the approximate formula $(HOCH_2NHC(O)NH)_nCH_2$.

Besides being more efficient in the utilization of the applied nutrients, slow-release technologies also reduce the impact on the environment and the contamination of the subsurface water. Slow-release fertilizers (various forms including fertilizer spikes, tabs, etc.) which reduce the problem of "burning" the plants due to excess nitrogen. Polymer coating of fertilizer ingredients gives tablets and spikes a 'true time-release' or 'staged nutrient release' (SNR) of fertilizer nutrients.

Controlled release fertilizers are traditional fertilizers encapsulated in a shell that degrades at a specified rate. Sulfur is a typical encapsulation material. Other coated products use thermoplastics (and sometimes ethylene-vinyl acetate and surfactants, etc.) to produce diffusion-controlled release of urea or other fertilizers.

"Reactive Layer Coating" can produce thinner, hence cheaper, membrane coatings by applying reactive monomers simultaneously to the soluble particles. "Multicote" is a process applying layers of low-cost fatty acid salts with a paraffin topcoat.

Foliar application

Foliar fertilizers are applied directly to leaves. The method is almost invariably used to apply water-soluble straight nitrogen fertilizers and used especially for high value crops such as fruits.

Chemicals that affect nitrogen uptake

Various chemicals are used to enhance the efficiency of nitrogen-based fertilizers. In this way farmers can limit the polluting effects of nitrogen run-off. Nitrification inhibitors (also known as nitrogen stabilizers) suppress the conversion of ammonia into nitrate, an anion that is more prone to leaching. 1-Carbamoyl-3-methylpyrazole (CMP), dicyandiamide, and nitrapyrin (2-chloro-6-trichloromethylpyridine) are popular. Urease inhibitors are used to slow the hydrolytic conversion of urea into ammonia, which is prone to evaporation as well as nitrification. The conversion of urea to ammonia catalyzed by enzymes called ureases. A popular inhibitor of ureases is N-(n-butyl)thiophosphoric triamide (NBPT).

Overfertilization

Fig. Fertilizer burn

Careful fertilization technologies are important because excess nutrients can be as detrimental. Fertilizer burn can occur when too much fertilizer is applied, resulting in drying out of the leaves and damage or even death of the plant. Fertilizers vary in their tendency to burn roughly in accordance with their salt index.

ENVIRONMENTAL EFFECTS

Fig. Runoff of soil and fertilizer during a rain storm

Fig. An algal bloom caused by eutrophication

WATER

Agricultural run-off is a major contributor to the eutrophication of fresh water bodies. For example, in the US, about half of all the lakes areeutrophic. The main contributor to eutrophication is phosphate, which is normally a limiting nutrient that allow the growth of algae, the demise of which consumes oxygen.

Furthermore, *alien* prokaryote specific amino acids (alien to all eukaryotes—plants, animals, true algae, amoeba, and fungi) produced bycyanobacteria blooms (commonly misnamed 'algal blooms'), such as microcystins toxins, can wreak havoc on eukaryote cellular machinery resulting in rapid death not only for the organism, but the eukaryote organisms that consume it. For example, cyanobacteria bloom toxins in waters absorbed by plants, which are consumed by fish, which is eaten by a deer, if humans drink the water, or eat any of the above—the plant, fish, or deer, the minute amounts (1 microgram per liter in water or 39 micrograms/kg in fish meat) of the toxins would be all that's required to pass on down and can cause acute liver failure in the human.

Boiling the microcystin tainted water will not destroy the toxins – it will increase the concentration of the toxins.—City of Toledo Water Advisory, August 2014

The nitrogen-rich compounds found in fertilizer runoff are the primary cause of serious oxygen depletion in many parts of oceans, especially in coastal zones, lakes and rivers. The resulting lack of dissolved oxygen greatly reduces the ability of these areas to sustain oceanic fauna. The number of oceanic dead zones near inhabited coastlines are increasing. As of 2006, the application of nitrogen fertilizer is being increasingly controlled in northwestern Europe and the United States. If eutrophication *can* be reversed, it may take decades before the accumulated nitrates in groundwater can be broken down by natural processes.

NITRATE POLLUTION

Only a fraction of the nitrogen-based fertilizers is converted to produce and other plant matter. The remainder accumulates in the soil or lost as run-off. High application rates of nitrogen-containing fertilizers combined with the high water-solubility of nitrate leads to increased runoff into surface water as well as leaching into groundwater. The excessive use of nitrogen-containing fertilizers (be they synthetic or natural) is particularly damaging, as much of the nitrogen that is not taken up by plants is transformed into nitrate which is easily leached.

Nitrate levels above 10 mg/L (10 ppm) in groundwater can cause 'blue baby syndrome' (acquired methemoglobinemia). The nutrients, especially nitrates, in fertilizers can cause problems for natural habitats and for human health if they are washed off soil into watercourses or leached through soil into groundwater.

SOIL

Acidulation

Also regular use of acidulated fertilizers generally contribute to the accumulation of soil acidity in soils which progressively increases aluminium availability and hence toxicity. The use of such acidulated fertilizers in the tropical and semi-tropical regions of Indonesia and Malaysia has contributed to soil degradation on a large scale from aluminium toxicity, which can only be countered by applications of limestone or preferably magnesian dolomite, which neutralises acid soil pH and also provides essential magnesium.

Nitrogen-containing fertilizers can cause soil acidification when added. This may lead to decreases in nutrient availability which may be offset by liming.

High levels of fertilizer may cause the breakdown of the symbiotic relationships between plant roots and mycorrhizal fungi.

HEAVY METAL ACCUMULATION

Cadmium

The concentration of cadmium in phosphorus-containing fertilizers varies considerably and can be problematic. Producers select phosphate rock based on the cadmium content. For example, mono-ammonium phosphate fertilizer may have a cadmium content of as low as 0.14 mg/kg or as high as 50.9 mg/kg. This is because the phosphate rock used in their manufacture can contain as much as 188 mg/kg cadmium (examples are deposits on Nauru and the Christmas islands). Continuous use of high-cadmium fertilizer can contaminate soil and plants. Limits to the cadmium content of phosphate fertilizersis has been considered by theEuropean Commission.

Fluoride

Phosphate rocks contain high levels of fluoride. Consequently the widespread use of phosphate fertilizers has increased soil fluoride concentrations. It has been found that food contamination from fertilizer is of little concern as plants accumulate little fluoride from the soil; of greater concern is the possibility of fluoride toxicity to livestock that ingest contaminated soils. Also of possible concern are the effects of fluoride on soil microorganisms.

Radioactive element accumulation

The radioactive content of the fertilizers varies considerably and depends both on their concentrations in the parent mineral and on the fertilizer production process.Uranium-238 concentrations range can range from 7 to 100 pCi/g in phosphate rock and from 1 to 67 pCi/g in phosphate fertilizers. Where high annual rates of phosphorus fertilizer are used, this can result in uranium-238 concentrations in soils and drainange waters that are several times greater than are normally present.However, the impact of these increases on the risk to human health from radinuclide contamination of foods is very small (less than 0.05 mSv/y).

Other metals

Steel industry wastes, recycled into fertilizers for their high levels of zinc (essential to plant growth), wastes can include the following toxic metals: lead arsenic, cadmium,chromium, and nickel. The most common toxic elements in this type of fertilizer are mercury, lead, and arsenic. Additionally phosphate fertilizers usually contain impurities some fluorides, cadmium, and uranium, although concentrations of the latter two heavy metals are dependent on the source of the phosphate and the fertilizer production process. These potentially harmful impurities can be removed; however, this significantly increases cost. Highly pure fertilizers are widely available and perhaps best known as the highly water soluble fertilizers containing blue dyes used around households. These highly water soluble fertilizers are used in the plant nursery business and are available in larger packages at significantly less cost than retail quantities. There are also some inexpensive retail granular garden fertilizers made with high purity ingredients.

Trace mineral depletion

Attention has been addressed to the decreasing concentrations of elements such as iron, zinc, copper and magnesium in many foods over the last 50–60 years. Intensive farming practices, including the use of synthetic fertilizers are frequently suggested as reasons for these declines and organic farming is often suggested as a solution.Although improved crop yields resulting from NPK fertilizers are known to dilute the concentrations of other nutrients in plants, much of the measured decline can be attributed to the use of progressively higher-yielding crop varieties which produce foods with lower mineral concentrations than their less productive ancestors. It is, therefore, unlikely that organic farming or reduced use of fertilizers will solve the problem; foods with high nutrient density are posited to be achieved using older, lower-yielding varieties or the development of new high-yield, nutrient-dense varieties.

Fertilizers are, in fact, more likely to solve trace mineral deficiency problems than cause them: In Western Australia deficiencies of zinc, copper, manganese, iron andmolybdenum were identified as limiting the growth of broad-acre crops and pastures in the 1940s and 1950s. Soils in Western Australia are very old, highly weathered and deficient in many of the major nutrients and trace elements. Since this time these trace elements are routinely added to fertilizers used in agriculture in this state. Many other soils around the world are deficient in zinc, leading to deficiency in both plants and humans, and zinc fertilizers are widely used to solve this problem.

ENERGY CONSUMPTION AND SUSTAINABILITY

In the USA in 2004, 317 billion cubic feet of natural gas were consumed in the industrial production of ammonia, less than 1.5% of total U.S. annual consumption of natural gas. A 2002 report suggested that the production of ammonia consumes about 5% of global natural gas consumption, which is somewhat under 2% of world energy production.

Ammonia is produced from natural gas and air. The cost of natural gas makes up about 90% of the cost of producing ammonia. The increase in price of natural gases over the past decade, along with other factors such as increasing demand, have contributed to an increase in fertilizer price.

Contribution to climate change

The greenhouse gases carbon dioxide, methane and nitrous oxide are produced during the manufacture of nitrogen fertilizer. The effects can be combined into an equivalent amount of carbon dioxide. The amount varies according to the efficiency of the process. The figure for the United Kingdom is over 2 kilogrammes of carbon dioxide equivalent for each kilogramme of ammonium nitrate. Nitrogen fertilizer can be converted by soil bacteria to nitrous oxide, a greenhouse gas.

ATMOSPHERE

Methane emissions from crop fields (notably rice paddy fields) are increased by the application of ammonium-based fertilizers. These emissions contribute to global climate change as methane is a potent greenhouse gas.

Through the increasing use of nitrogen fertilizer, which was used at a rate of about 110 million tons (of N) per year in 2012, adding to the already existing amount of reactive nitrogen, nitrous oxide (N_2O) has become the third most important greenhouse gas after carbon dioxide and methane. It has a global warming potential 296 times larger than an equal mass of carbon dioxide and it also contributes to stratospheric ozone depletion. By changing processes and procedures, it is possible to mitigate some, but not all, of these effects on anthropogenic climate change.

REGULATION

In Europe problems with high nitrate concentrations in run-off are being addressed by the European Union's Nitrates Directive. Within Britain, farmers are encouraged to manage their land more sustainably in 'catchment-sensitive farming'. In the US, high concentrations of nitrate and phosphorus in runoff and drainage water are classified as non-point source pollutants due to their diffuse origin; this pollution is regulated at state level. Oregon and Washington, both in the United States, have fertilizer registration programs with on-line databases listing chemical analyses of fertilizers.

HISTORY

Management of soil fertility has been the preoccupation of farmers for thousands of years. Egyptians, Romans, Babylonians, and early Germans all are recorded as using minerals and or manure to enhance the productivity of their farms. The modern science of plant nutrition started in the 19th century and the work of German chemist Justus von Liebig, among others. John Bennet Lawes, an Englishentrepreneur, began to experiment on the effects of various manures on plants growing in pots in 1837, and a year or two later the experiments were extended to crops in the field. One immediate consequence was that in 1842 he patented a manure formed by treating phosphates with sulphuric acid, and thus was the first to create the artificial manure industry. In the succeeding year he enlisted the services of Joseph Henry Gilbert, with whom he carried on for more than half a century on experiments in raising crops at the Institute of Arable Crops Research.

The Birkeland–Eyde process was one of the competing industrial processes in the beginning of nitrogen based fertilizer production.This process was used to fix atmospheric nitrogen (N_2) into nitric acid (HNO_3), one of several chemical processes generally referred to as nitrogen fixation. The resultant nitric acid was then used as a source of nitrate (NO_3^-). A factory based on the process was built in Rjukan and Notodden in Norway, combined with the building of large hydroelectric power facilities.

The 1910s and 1920s witness the rise of the Haber process and the Ostwald process. The Haber process produces ammonia (NH_3) from methane (CH_4) gas and molecular nitrogen (N_2). The ammonia from the Haber process is then converted into nitric acid (HNO_3) in the Ostwald process. The development of synthetic fertilizer has significantly supported global population growth — it has been estimated that almost half the people on the Earth are currently fed as a result of synthetic nitrogen fertilizer use.

The use of commercial fertilizers has increased steadily in the last 50 years, rising almost 20-fold to the current rate of 100 million tonnes of nitrogen per year. Without commercial fertilizers it is estimated that about one-third of the food produced now could not be produced. The use of phosphate fertilizers has also increased from 9 million tonnes per year in 1960 to 40 million tonnes per year in 2000. A maize crop yielding 6–9 tonnes of grain per hectare requires 31–50 kg of phosphate fertilizer to be applied, soybean requires 20–25 kg per hectare. Yara International is the world's largest producer of nitrogen based fertilizers.

Controlled-nitrogen-release technologies based on polymers derived from combining urea and formaldehyde were first produced in 1936 and commercialized in 1955. The early product had 60 percent of the total nitrogen cold-water-insoluble, and the unreacted (quick release) less than 15%. Methylene ureas were commercialized in the 1960s and 1970s, having 25 and 60% of the nitrogen cold-water-insoluble, and unreacted urea nitrogen in the range of 15 to 30%.

In the 1960s, the National Fertilizer Development Centre began developing Sulfur-coated urea; sulfur was used as the principle coating material because of its low cost and its value as a secondary nutrient. Usually there is another wax or polymer which seals the sulfur; the slow release properties depend on the degradation of the secondary sealant by soil microbes as well as mechanical imperfections (cracks, etc.) in the sulfur. They typically provide 6 to 16 weeks of delayed release in turf applications. When a hard polymer is used as the secondary coating, the properties are a cross between diffusion-controlled particles and traditional sulfur-coated.

DEFINITION OF FERTILIZERS

Organic or inorganic plant foods, which may be either liquid or granular, used to amend the soil in order to improve the quality or quantity of plant growth. Fertilizers or fertilizers are compounds given to plants with the intention of promoting growth; they are usually applied either via the soil, for uptake by plant roots, or by foliar spraying, for uptake through leaves. Fertilizers can be organic (composed of organic matter, *i.e.*, carbon based), or inorganic (containing simple, inorganic chemicals).

They can be naturally-occurring compounds such as peat or mineral deposits, or manufactured through natural processes or chemical processes (such as the Haber process).

Fertilizers typically provide, in varying proportions, the three major plant nutrients (nitrogen, phosphorus, and potassium), the secondary plant nutrients (calcium, sulphur, magnesium), and sometimes trace elements (or micronutrients) with a role in plant nutrition: boron, manganese, iron, zinc, copper and molybdenum.

INORGANIC FERTILIZERS

- Examples of naturally-occurring inorganic fertilizers include diatomaceous earth and limestone.
- Examples of manufactured or chemically-synthesized inorganic fertilizers include ammonium nitrate, potassium sulphate, and superphosphate, or triple superphosphate.

Synthesised materials are also called artificial fertilizers, and may be described as straight, where the product predominantly contains the three primary ingredients of nitrogen (N), phosphorus (P) and potassium/potash (K), often described as NPK fertilizers. They are named or labelled according to the content of these three elements, thus a 5-10-5 fertilizer would have 10 per cent phosphate in its ingredients. If nitrogen is the main element, they are often described as nitrogen fertilizers.

Alternatively they may be described as compound where there is a mix of nutrients. Chemist Justus von Liebig (in the 19th century) contributed greatly to understanding the role of inorganic compounds in plant nutrition and devised the concept of Liebig's barrel to illustrate the significance of inadequate concentrations of essential nutrients. At the same time he deemphasised the role of humus. This theory was influential in the great expansion in use of artificial fertilizers in the 20th century.

Nitrogen fertilizer is often synthesised using the Haber-Bosch process, which produces ammonia. This ammonia is applied directly to the soil or used to produce other compounds, notably ammonium nitrate, a dry, concentrated product. It can also be used in the Odda Process to produce compound fertilizers such as 15-15-15.

Inorganic fertilizers sometimes do not replace trace mineral elements in the soil which become gradually depleted by crops grown there. This has been linked to studies which have shown a marked fall (up to 75 per cent) in the quantities of such minerals present in fruit and vegetables.

One exception to this is in Western Australia where deficiencies of zinc, copper, manganese, iron and molybdenum were identifed as limiting the growth of crops and pastures in the 1940's and 1950's. Soils in Western Australia are very old, highly weathered and deficient in many of the major nutrients and trace elements. Since, this time these trace elements are routinely added to inorganic fertilizers used in Agriculture in this state. In many countries there is the public perception that inorganic fertilizers 'poison the soil' and result in "low quality" produce.

However, there is very little (if any) scientific evidence to support these views. When used appropriately, inorganic fertilizers enhance plant growth, the accumulation of organic matter and the biological activity of the soil, while reducing the risk of water run-off, overgrasing and soil erosion. The nutritional value of plants for human and animal consumption is typically improved when inorganic fertilizers are used appropriately.

ORGANIC FERTILIZERS

- Examples of naturally occurring organic fertilizers include manure and slurry, urine, peat, seaweed and guano. Green manure crops are also grown to add nutrients to the soil. Naturally occurring minerals such as mine rock phosphate, sulphate of potash and limestone are also considered Organic Fertilizers.

- Examples of manufactured organic fertilizers include compost, dried blood, bone meal and seaweed extracts. Other examples are natural enzyme digested proteins, fish meal, and feather meal. A listing of products may be found at Organic Materials Review Institute and California Organic Fertilizers Inc..

The decomposing crop residue from prior years is another source of fertility. Though not strictly considered "fertilizer", the distinction seems more a matter of words than reality.

Some ambiguity in the usage of the term 'organic' exists because some of synthetic fertilizers, such as urea and urea formaldehyde, are fully organic in the sense of organic chemistry. In fact, it would be difficult to chemically distinguish between urea of biological origin and that produced synthetically. On the other hand, some fertilizer materials commonly approved for organic agriculture, such as powdered limestone, diatomaceous earth, and Chilean saltpeter, are inorganic in the use of the term by chemistry.

Although the density of nutrients in organic material is comparatively modest, they have some advantages. For one thing organic growers typically produce some or all of their fertilizer on-site, thus lowering operating costs considerably.

Then there is the matter of how effective they are at promoting plant growth, chemical soil test results aside. The answers are encouraging. Since, the majority of nitrogen supplying organic fertilizers contain insoluble nitrogen and are slow release fertilizers their effectiveness can be greater than conventional nitrogen fertilzers.

Implicit in modern theories of organic agriculture is the idea that the pendulum has swung the other way to some extent in thinking about plant nutrition. While admitting the obvious success of Leibig's theory, they stress that there are serious limitations to the current methods of implementing it via chemical fertilization.

They re-emphasise the role of humus and other organic components of soil, which are believed to play several important roles:

- Mobilising existing soil nutrients, so that good growth is achieved with lower nutrient densities while wasting less.
- Releasing nutrients at a slower, more consistent rate, helping to avoid a boom-and-bust pattern.
- Helping to retain soil moisture, reducing the stress due to temporary moisture stress.
- Improving the soil structure.

Organics also have the advantage of avoiding certain long-term problems associated with the regular heavy use of artificial fertilizers;

- The possibility of 'burning' plants with the concentrated chemicals (*i.e.*, an over supply of some nutrients).
- The progressive decrease of real or perceived "soil health", apparent in loss of structure, reduced ability to absorb precipitation, lightening of soil colour, etc.
- The necessity of reapplying artificial fertilizers regularly (and perhaps in increasing quantities) to maintain fertility.
- The cost (substantial and rising in recent years) and resulting lack of independence.

Organic fertilizers also have their disadvantages. They are typically a dilute source of nutrients compared to inorganic fertilizers, and where significant amounts of nutrients are required for profitable yields, very large amounts of organic fertilizers must be applied. This results in prohibitive transportation and application costs, especially where the agriculture is practiced a long distance from the source of the organic fertilizer.

The composition of organic fertilizers tends to be highly variable, so that accurate application of nutrients to match plant production is difficult. Hence, large-scale agriculture tends to rely on inorganic fertilizers while organic fertilizers are cost-effective on small-scale horticultural or domestic gardens. Finally, some organic fertilizers such as manures can contain bacteria or heavy metals harmful to human health.

In practice a compromise between the use of artificial and organic fertilizers is common, typically by using inorganic fertilizers supplemented with the application of organics that are readily available such as the return of crop residues or the application of manure.

It is important to differentiate between what we mean by organic fertilizers and fertilizers approved for use in organic farming and organic gardening by organisations and authorities who provide organic certification services. Some approved fertilizers may be inorganic, naturally occurring chemical compounds, *e.g.*, minerals.

ORGANIC FERTILIZERORGANIC FERTILIZER AND ORGANIC MATTER

Organic fertilizers are fertilizer compounds that contain one or more kinds of organic matter. The ingredients may be animal or vegetable matter or a combination of the two. It is possible to purchase commercial brands of organic rich fertilizer as well as prepare organic fertilizer at home by building a compost heap.

When describing fertilizer, it is important to remember that the working definition of organic is the same as that applied to organic foods. That is, the fertilizer is composed of elements that are produced in a completely natural manner, without the aid of any synthetically manufactured components or additives.

Many different natural elements can go into the creation of organic fertilizer. Animal manure is a common ingredient in both commercial and home prepared blends. Rotten produce, bone meal, and the decomposing plants removed at the end of the season can also be chopped or ground in to small particles for inclusion in the fertilizer. Essentially, any matter that is of natural origin and subject to decomposition is a good candidate for inclusion in the product.

Organically prepared fertilizing agents can be used to grow vegetables, raise flowers, and even to produce a lush green lawn. As with any type of fertilizing product, the organic grass fertilizer is spread evenly across the expanse of the yard. Often, the fertilization process takes place before the grass seeds are sown. Once the soil is properly tilled and mixed with the organic lawn fertilizer, the seeds are distributed and the area is watered. The presence of the natural materials helps the seeds to spout and take root, eventually producing a beautiful carpet of grass across the lawn.

In like manner organic fertilizer can be used in vegetable gardens to replenish the nutrient content in the soil. This is often done prior to planing the next round of crops in the garden. Preparing the soil in advance provides the ideal setting for the newly planted seeds or young plants to take root and grow rapidly.

Even flower gardens can benefit from the use of organic fertilizer. As with lawns and vegetable gardens, the fertilizer is added to the soil prior to planting. Depending on the climate and the types of flowers in the garden, the use of organic liquid fertilizer may also be recommended as the plants mature. An organic nitrogen fertilizer may also be helpful during the growth process for both flowers and lawns.

Organic fertilizer can be purchased in most farmers' exchange stores, along with many home and garden shops. Since the natural fertilizer is usually sold alongside brands that contain synthetic elements, it is important to read the contents listed on the bag of fertilizer before making the purchase.

DISCUSSION OF THE TERM'ORGANIC'

There can be confusion as to the definition of the word'organic' as applied to the agricultural systems and fertilizer. The problem is related to the colloquial versus technical usage of the term.

NATURAL SOURCING

Animal-sourced Urea and Urea-Formaldehyde are suitable for application organic agriculture, while pure synthetic forms are not deemed, however, pure urea is not. The common thread that can be seen through these examples is that organic agriculture attempts to define itself through minimal processing as well as being naturally-occurring or via natural biological processes such as composting. Cover crops are also grown to enrich soil as a green manure through nitrogen fixation from the atmosphere; as well as phosphorus content of soils. Powdered limestone, mined rock phosphate and Chilean saltpeter, are inorganic chemicals in the technical sense of the word, but are considered suitable for organic agriculture.

ADVANTAGES

Although the density of nutrients in organic material is comparatively modest, they have many advantages. The majority of nitrogen supplying organic fertilizers contain insoluble nitrogen and act as a slow-release fertilizer. By their nature, organic fertilizers increase physical and biological nutrient storage mechanisms in soils, mitigating risks of over-fertilization. Organic fertilizer nutrient content, solubility, and nutrient release rates are typically much lower than mineral fertilizers A University of North Carolina study found that potential mineralizable nitrogen (PMN) in the soil was 182-285% higher in organic mulched systems, than in the synthetics control.

Organic fertilizers also have the advantage of avoiding certain problems associated with the regular heavy use of artificial fertilizers:

- The necessity of reapplying artificial fertilizers regularly to maintain fertility
- Extensive runoff of soluble nitrogen and phosphorus, leading to eutrophication of bodies of water
- Costs are lower for if fertilizer is locally available

It is widely thought that organic fertilizer is better than inorganic fertilizer. However, balanced responsible use of either or both can be just as good for the soil.

DISADVANTAGES

Organic fertilizers have the following disadvantages:

- As a dilute source of nutrients when compared to inorganic fertilizers, transporting large amount of fertilizer will incur higher costs. Especially with slurry and manure
- The composition of organic fertilizers tends to be more complex and variable than a standardized inorganic product.
- Improperly-processed organic fertilizers may contain pathogens from plant or animal matter that are harmful to humans or plants. However, proper composting should remove them.

CONVENTIONAL FARMING APPLICATION

In non-organic farming a compromise between the use of artificial and organic fertilizers is common, often using inorganic fertilizers supplemented with the application of organics that are readily available such as the return of crop residues or the application of manure.

FEATHER MEAL

Hydrolyzed poultry feathers or feather meal is produced by hydrolyzing clean, undecomposed feathers from slaughtered poultry.

The most important factor affecting the quality of hydrolyzed poultry feathers is the extent of hydrolyzation. If less than 75 per cent of the crude protein content is digestible by the pepsin digestibility method, then hydrolyzation was incomplete and protein quality is reduced.

The protein in feather meal is degraded slowly in the rumen compared to most other protein sources. In research at Purdue University, a combination of feather meal and urea produced average daily gains in growing beef cattle similar to that achieved with soybean meal.

There is very little experience with feather meal in dairy cattle rations. It is not very palatable and should be introduced into the ration gradually. It may be fed to milking cows at an average rate of up to 1 ½ pounds (0.7 kg) per cow per day.

Uses

- Used to increase green leaf growth
- Can be used as a compost decomposition activator
- Improves soil structure
- When adding to your garden as a nitrogen source, always blend it into the soil

Organic garden fertilizers are non-synthetic and non-petroleum chemical based naturally occurring substances used to make soil more fertile. The following organic garden fertilizers are available commercially. Organic garden fertilizers should only be used to supplement soil fertility and not as the sole source of your soil fertility program. Its important to supplement with organic fertilizers with compost and cover crops as your primary source of additional nutrients.

BACKGROUND OF ORGANIC FERTILIZER

Fig. A cement reservoir containing cow manure mixed with water. This is common in rural Hainan Province, China. Note the bucket on a stick that the farmer uses to apply the mixture.

Organic fertilizers are fertilizers derived from animal matter or vegetable matter. (e.g. compost, manure). In contrast, the majority of fertilizers are extracted from minerals (e.g., phosphate rock) or produced industrially (e.g., ammonia). Naturally occurring organic matter|organic fertilizers include animal wastes from meat processing, peat, manure, slurry, and guano.

EXAMPLES AND SOURCES OF ORGANIC FERTILIZER

The main organic fertilizers are in ranked order: peat, animal wastes (often from slaughter houses), plant wastes from agriculture, and sewage sludge.

MINERAL

The main source of organic fertilizer is peat, an immature precursor to coal. Peat itself offers no nutritional value to the plants, but improves the soil by aeration and absorbing water.

Fig. Peat is the most widely used organic fertilizer.

Mined powdered limestone, rock phosphate, and Chilean saltpeter are inorganic (not of biologic origins) compounds, which can be energetically intensive to harvest.

ANIMAL SOURCES

These materials include the products of the slaughter of animals. Bloodmeal, bone meal, hides, hoofs, and horns are typical precursors.

Chicken litter, which consists of chicken manure mixed with sawdust, is an organic fertilizer that has been shown to better condition soil for harvest than synthesized fertilizer. Researchers at the Agricultural Research Service (ARS) studied the effects of using chicken litter, an organic fertilizer, versus synthetic fertilizers on cotton fields, and found that fields fertilized with chicken litter had a 12% increase in cotton yields over fields fertilized with synthetic fertilizer. In addition to higher yields, researchers valued commercially sold chicken litter at a $17/ton premium (to a total valuation of $78/ton) over the traditional valuations of $61/ton due to value added as a soil conditioner.

PLANT

Processed organic fertilizers include compost, humic acid, amino acids, and seaweed extracts. Other examples are natural enzyme-digested proteins, fish meal, and feather meal. Decomposing crop residue (green manure) from prior years is another source of fertility.

Other ARS studies have found that algae used to capture nitrogen and phosphorus runoff from agricultural fields can not only prevent water contamination of these nutrients, but also can be used as an organic fertilizer. ARS scientists originally developed the "algal turf scrubber" to reduce nutrient runoff and increase quality of water flowing into streams, rivers, and lakes. They found that this nutrient-rich algae, once dried, can be applied to cucumber and corn seedlings and result in growth comparable to that seen using synthetic fertilizers.

SEWAGE SLUDGE

Although night soil is a traditional organic fertilizer, the main source of this type is sewage sludge.

Fig. Decomposing animal manure, an organic fertilizer source

Recycled sewage sludge (aka biosolids) as soil amendment is only available to less than 1% of US agricultural land. Industrial pollutants in sewage sludge prevents recycling it as fertilizer. The USDA prohibits use of sewage sludge in organic agricultural operations in the U.S. due to industrial pollution, pharmaceuticals, hormones, heavy metals, and other factors. The USDA now requires 3rd-party certification of high-nitrogen liquid organic fertilizers sold in the U.S.

Sewage sludge use in organic agricultural operations in the U.S. has been extremely limited and rare due to USDA prohibition of the practice (due to toxic metal accumulation, among other factors).

Animal sourced urea and urea-formaldehyde from urine are suitable for organic agriculture; however, synthetically produced urea is not. The common thread that can be seen through these examples is that *organic* agriculture attempts to define itself through minimal processing, as well as being naturally occurring or via natural biological processes such as composting.

CHEMICAL VS. ORGANIC

Many organic materials serve as both fertilizers and soil conditioners—they feed both soils and plants. This is one of the most important differences between a chemical approach and an organic approach toward soil care and fertilizing. Soluble chemical fertilizers contain mineral salts that plant roots can absorb quickly. However, these salts do not provide a food source for soil microorganisms and earthworms, and will even repel earthworms because they acidify the soil. Over time, soils treated only with synthetic chemical fertilizers lose organic matter and the all-important living organisms that help to build a quality soil. As soil structure declines and water-holding capacity diminishes, more and more of the chemical fertilizer applied will leach through the soil. In turn, it will take ever-increasing amounts of chemicals to stimulate plant growth. When you use organic fertilizers, you avoid throwing your soil into this kind of crisis condition.

The manufacturing process of most chemical fertilizers depends on nonrenewable resources, such as coal and natural gas. Others are made by treating rock minerals with acids to make them more soluble. Fortunately, there are more and more truly organic fertilizers coming on the market. These products are made from natural plant and animal materials or from mined rock minerals. However, the national standards that define and distinguish organic fertilizers from chemical fertilizers are complicated, so it's hard to be sure that a commercial fertilizer product labeled "organic" truly contains only safe, natural ingredients. Look for products labeled "natural organic," "slow release," and "low analysis." Be wary of products labeled organic that have an NPK (nitrogen-phosphorus-potassium) ratio that adds up to more than 15. Ask a reputable garden center owner to recommend fertilizer brands that meet organic standards.

USING ORGANIC FERTILIZERS

If you're a gardener who's making the switch from chemical to organic fertilizers, you may be afraid that using organic materials will be more complicated and less convenient than using premixed chemical fertilizers. Not so! Organic fertilizer blends can be just as convenient and effective as blended synthetic fertilizers. You don't need to custom feed your plants organically unless it's an activity you enjoy. So while some experts will spread a little blood meal around their tomatoes at planting, and then some bonemeal just when the blossoms are about to pop, most gardeners will be satisfied to make one or two applications of general-purpose organic fertilizer throughout the garden.

Convenient products like dehydrated organic cow-manure pellets and liquid seaweed make it easy to fertilize houseplants and containers too. (Don't use fish emulsion indoors, though, because of its strong odor. Save it for your outdoor containers and garden plants.)

If you want to try a plant-specific approach to fertilizing, you can use a variety of specialty organic fertilizers that are available from mail-order 231 supply companies or at most well-stocked garden centers and home centers. You can find everything from organic tomato and rose fertilizer mixes to organic fertilizer mixes especially created for transplants, lawns, heavy bloom production, even containers.

You can also make custom mixes to address your plants' specific needs. For example, you can use bat and bird guano, composted chicken manure, blood meal, chicken-feather meal, or fish meal as nitrogen sources. Bonemeal is a good source of phosphorus, and kelp or greensand are organic sources of potassium.

DRY ORGANIC FERTILIZERS

Dry organic fertilizers can consist of a single material, such as rock phosphate or kelp (a type of nutrient-rich seaweed), or they can be a blend of many ingredients. Almost all organic fertilizers provide a broad array of nutrients, but blends are specially formulated to provide balanced amounts of nitrogen, potassium, and phosphorus, as well as micronutrients. There are several commercial blends, but you can make your own general-purpose fertilizer by mixing individual amendments.

APPLYING DRY FERTILIZERS

The most common way to apply dry fertilizer is to broadcast it and then hoe or rake it into the top 4 to 6 inches of soil. You can add small amounts to planting holes or rows as you plant seeds or transplants. Unlike dry synthetic fertilizers, most organic fertilizers are nonburning and will not harm delicate seedling roots.

During the growing season, boost plant growth by side-dressing dry fertilizers in crop rows or around the drip line of trees or shrubs. It's best to work side-dressings into the top inch of the soil.

LIQUID ORGANIC FERTILIZERS

Use liquid fertilizers to give your plants a light nutrient boost or snack every month or even every 2 weeks during the growing season. Simply mix the foliar spray in the tank of a backpack sprayer, and spray all your plants at the same time.

Plants can absorb liquid fertilizers through both their roots and through leaf pores. Foliar feeding can supply nutrients when they are lacking or unavailable in the soil, or when roots are stressed. It is especially effective for giving fast-growing plants like vegetables an extra boost during the growing season. Some foliar fertilizers, such as liquid seaweed (kelp), are rich in micronutrients and growth hormones. These foliar sprays also appear to act as catalysts, increasing nutrient uptake by plants. Compost tea and seaweed extract are two common examples of organic foliar fertilizers.

APPLYING LIQUID FERTILIZERS

With flowering and fruiting plants, foliar sprays are most useful during critical periods (such as after transplanting or during fruit set) or periods of drought or extreme temperatures. For leaf crops, some suppliers recommend biweekly spraying.

When using liquid fertilizers, always follow label instructions for proper dilution and application methods. You can use a surfactant, such as coconut oil or a mild soap (¼ teaspoon per gallon of spray), to ensure better coverage of the leaves. Otherwise, the spray may bead up on the foliage and you won't get maximum benefit. Measure the surfactant carefully; if you use too much, it may damage plants. A slightly acid spray mixture is most effective, so check your spray's pH. Use small amounts of vinegar to lower pH and baking soda to raise it. Aim for a pH of 6.0 to 6.5.

Any sprayer or mister will work, from hand-trigger units to knapsack sprayers. Set your sprayer to emit as fine a spray as possible. Never use a sprayer that has been used to apply herbicides.

The best times to spray are early morning and early evening, when the liquids will be absorbed most quickly and won't burn foliage. Choose a day when no rain is forecast and temperatures aren't extreme.

Spray until the liquid drips off the leaves. Concentrate the spray on leaf undersides, where leaf pores are more likely to be open. You can also water in liquid fertilizers around the root zone. Adrip irrigation system can carry liquid fertilizers to your plants. Kelp is a better product for this use, as fish emulsion can clog the irrigation emitters.

USING GROWTH ENHANCERS

Growth enhancers are materials that help plants absorb nutrients more effectively from the soil. The most common growth enhancer is kelp (a type of seaweed), which has been used by farmers for centuries.

Kelp is sold as a dried meal or as an extract of the meal in liquid or powdered form. It is totally safe and provides some 60 trace elements that plants need in very small quantities. It also contains growth-promoting hormones and enzymes. These compounds are still not fully understood, but are involved in improving a plant's growing conditions.

APPLYING GROWTH ENHANCERS

Follow the directions for spraying liquid fertilizers when applying growth enhancers as a foliar spray.

You can also apply kelp extract or meal directly to the soil; soil application will stimulate soil bacteria. This in turn increases fertility through humus formation, aeration, and moisture retention.

Apply 1 to 2 pounds of kelp meal per 100 square feet of garden each spring. Apply kelp extract once a month for the first 4 or 5 months of the growing season.

If fresh seaweed is available, rinse it to remove the sea salt and spread it over the soil surface in your garden as a mulch, or compost it. Seaweed decays readily because it contains little cellulose.

2

Phosphatic Fertilizers

INTRODUCTION

Phosphate fertilizer is a fertilizer that is high in phosphorous. Most phosphate fertilizer comes from phosphate rock, a mineral mined in massive quantities of millions of tons from locations around the world. This mineral provides one of the three main nutrients needed by all plants for vigorous growth; the other two are nitrogen and potassium.

Phosphate rock is most often mined in large open pit mines; notable deposits are in Morocco, China, Florida and South America. This raw ore is occasionally used as a fertilizer without any further processing, especially in acid soils, where it serves the dual purpose of raising pH. Rock phosphate fertilizers are becoming less common, as the raw rock provides relatively little phosphorous for its weight and transportation costs make it more expensive than refined phosphate fertilizer.

The process by which phosphate rock is converted to phosphate fertilizer involves treatment with sulfuric acid; the result is often called "super phosphate." This treatment with sulfuric acid draws the phosphates from the raw ore and creates a water soluble form. This is mixed with water to create a number of similar fertilizer compounds in a concentrated liquid form which is easy to apply to fields and crops.

MANUFACTURING PHOSPHATE FERTILIZER

Phosphate fertilizer is manufactured from rock phosphate. Rock phosphate is mined and processed into phosphate fertilizer primarily in the United States, North Africa, and the former Soviet Union. Historically, rock phosphate was used as a fertilizer, but because of low concentration and low availability, most applied fertilizer has been processed into a higher concentration.

TYPES AND CONCENTRATIONS

There are several types of fertilizer available on the market, in both liquid and dry granular form.

The choice of liquid or dry formulation should be based on the fit for individual farming operations. The effects of environmental conditions, soil properties, and method of application are the same for liquid and dry formulations. The fertilizer choice should be based on cost, availability, and fertilizer placement options.

Similarly, soil pH should not influence the choice in phosphate fertilizer formulations. Agronomic studies have not found differences between MAP and DAP, so formulation selection should be based on other factors.

P fertilizer type	Analysis	Properties
Rock phosphate	0-(3 to 8)-0	Raw mined product with limited phosphate availability
Superphosphate (OSP or SSP)	0-20-0-10S	Results from reaction of rock P with sulfuric acid - an older form of phosphate
Triple Super Phosphate (monocalcium phosphate)	0-45-0	Solid granular fertilizer with good handling properties
Monoammonium phosphate (MAP)	11-48-0 11-51-0 11-52-0 11-54-0 11-55-0	A solid granular that does not absorb moisture during storage. It is fairly resistant to breakdown during handling
Diammonium phosphate (DAP)	18-46-0	Solid granular fertilizer with higher level of nitrogen
Ammonium Polyphosphate	10-34-0	The solution will settle out below -18 °C and must be brought back into solution

Many dry blends of urea and ammonium phosphate are also available, with analysis such as 28-28-0, 17-34-0, etc.

POLOYPHOSPHATE VS. ORTHOPHOSPHATE

During fertilizer production, rock phosphate reacts with an acid to produce phosphoric acid. Two types of phosphorous compounds are present. Polyphosphates are a series of orthophosphates that are chemically joined together. Polyphosphate is basically two or more orthophosphates ions combined together when water is removed.

Some phosphate fertilizer contain orthophosphates only, while others have both poly- and orthophosphates. However, upon contact with the soil, polyphosphates will absorb water and readily convert back to orthophosphate. This conversion can be completed in days with normal soil temperatures, so crop response to phosphate fertilizer is the same, no matter what the formulation base.

The effect of orthophosphate and polyphosphate fertilizers on crop production was evaluated at the University of Minnesota at 15, 30, and 45 lb/ac of applied P_2O_5. Resulting yields at the three levels of applied fertilizer were the same for both poly- and orthophosphate. Note: 11-52-0 is 100% orthophosphate, most liquid fertilizers are polyphosphates.

THE INFLUENCE OF P SOURCE ON CORN YIELD

P_2O_5 Applied lb/acre	Polyphosphate bu/acre	P Source Orthophosphate
15	124	124
30	134	134
45	142	142

CITRIC ACID SOLUBLE PHOSPHATIC FERTILIZERS

Citric acid soluble phosphatic fertilizers are not soluble in water but are readily soluble in acidic water or weak acids like 2 per cent citric acid. They also contain phosphorus in available form, i.e., HPO_4. The fertilizers are suitable for acidic soils where they can easily dissolve and become available to plants. The examples of these fertilziers are:

a) Basic slag -18% phosphate (P_2OS)

b) Dicalcium phosphate -34-39% P20S

c) Rhenania phosphate -23-26% P20S

The first two are very important and are commonly used in a few pockets of the country.

a. Basic slag: Basic slag is a by-product of iron and steel industries. Original iron ore used as a raw material in iron industries contains appreciable amount of phosphorus as impurities. Purer form of iron is exracted from raw material leaving phosphorus and calcium rich basic slag as a by-product. Basic slag produced in India is of low quality. In contains only 3-8 per cent phosphorus in comparison to 14-18% phosphorus in basic slag produced in European countries. This is a grayish black powder with a very high specific weight., Its phosphorus is easily soluble in soil water of acidic soils. It exerts alkaline residual effect in the soil, therefore is useful for applying in acid soils. Basic slag is also used as liming material in our country.

b. Dicalcium phosphate: Dicalcium phosphate has a excellent physical condition. It is also very rich in phosphorus (34% citrate soluble phosphates), therefore, less costly than basic slag in transport, storage and distribution. This fertilizer is suitable for a wide range of crops and soils because its phosphorus does not so easily convert into unavailable form as that of super phosphates. Dicalcium phosphate is suitable for acidic soils.

c. Renonia phosphate: This is a citrate solube phosphatic fertilizer which contains alteast 24% available phosphorus (HPO_4)' The production and consumption of this ferrtilizer in our country is negligible. However, in foreign countries this fertilizer is used in large amounts.

SOIL FERTILITY MANAGEMENT

FERTILIZER MATERIALS

Phosphorus fertilizers

Plants absorb most of their phosphorus from the soil solution as orthophosphate (H_2PO_4-), regardless of the original source of phosphorus. Although orthophosphate's negative charge prevents it from being attracted by the soil's cation exchange capacity (CEC), it does react strongly in the soil, primarily with the large amount of iron and aluminum naturally in the soil, to form products that are very insoluble and thus unavailable to plants. A major factor controlling these reactions is the soil pH. At low or high pH, the solubility of phosphorus (and thus its availability) is very low. The maximum availability occurs in the 6.0 to 7.0 pH range. This is another important reason to lime regularly.

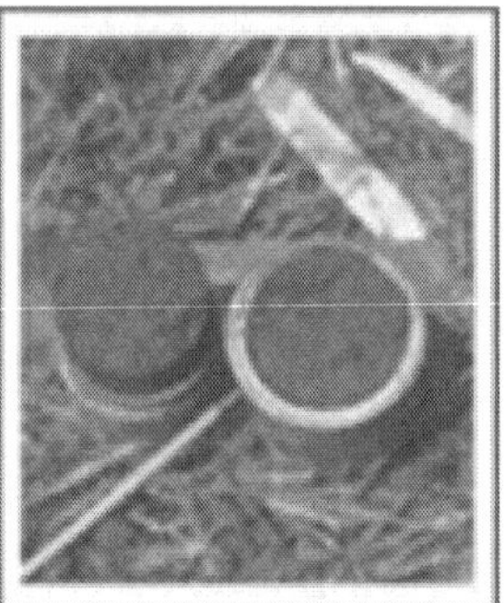

The solubility of phosphorus in fertilizer varies. The legal definition of available phosphorus in fertilizer is the sum of the phosphorus that is soluble in water plus that which is soluble in a citrate solution. Regardless of the actual chemical form of the phosphorus, the analyses of phosphorus fertilizers are given as phosphate (P_2O_5). The water solubility of this phosphorus can vary from 0 to 100 percent. Generally, the higher the water solubility, the more effective the phosphorus source. This is especially important for shortseason, fast-growing crops, for crops with restricted root systems, for starter fertilizers, and for areas where less-than-optimum rates of phosphorus are applied to soils testing low in phosphorus.

The most common phosphate fertilizers are triple superphosphate (0–46–0), monoammonium phosphate (11–52–0), diammonium phosphate (18–46–0), and ammonium polyphosphate (10–34–0) liquid. All of these materials are highly water soluble. The ammonium phosphates also are excellent nitrogen sources. Monoammonium phosphate and ammonium polyphosphate, either alone or with some added potassium, make excellent starter fertilizers because of their high P-to-N ratios, high water solubility, and low free ammonia. Diammonium phosphate (DAP) is not recommended as a starter material because it produces free ammonia, which can harm the seed. Many starter fertilizers, however, contain DAP; thus it is critical, when that is the case, that the starter is accurately placed a safe distance (about 2 inches) from the seed, and that high rates are avoided. Because P is relatively immobile, placement where plant roots will have easy access to the fertilizer is especially important for P fertilizers.

POTASSIUM FERTILIZERS

Potassium occurs in the soil in three forms: as exchangeable (available) potassium (K^+) adsorbed onto the soil CEC; fixed by certain minerals from which it is released very slowly to available form; and in unavailable mineral forms (most of the potassium in soils). Plants take up potassium as the K^+ ion.

The common source of fertilizer potassium is muriate of potash (0–0–60), which chemically is potassium chloride (KCl). Potassium chloride is highly water soluble. At excessive rates, muriate of potash can cause salt damage to plants.

BLENDED FERTILIZERS

Blended fertilizers are made by physically mixing fertilizer materials to give a desired grade. The individual particles remain separate in the mixture, and segregation may occur. This problem can be reduced by using materials with the same particle size. The quality of the blended fertilizer, particularly the segregation of the individual fertilizer materials, should be an important factor in choosing a fertilizer. Poor quality blends may contain the guaranteed nutrients but if there is segregation these nutrients will not be supplied uniformly to the crop. If properly made so as to reduce segregation during transportation and application, blends generally are equal in agronomic effectiveness to granulated complete fertilizers. Blends have the added advantage of allowing a very wide range of fertilizer grades, thus making it possible to match a fertilizer exactly to a soil test recommendation.

When using a blend as a starter fertilizer, try to avoid urea and diammonium phosphate as ingredients. Both materials produce free ammonia, which can hurt seed germination and seedling growth.

FLUID FERTILIZERS

Fluid fertilizers are becoming more and more common. Recent fertilizer statistics for Pennsylvania indicate that nitrogen solution (UAN) is the most common source of nitrogen used in the state. Multinutrient fluid fertilizers also are becoming more popular. The fluid fertilizers may be categorized into two groups: clear solutions and suspensions.

In clear solutions, which are the most common fluid fertilizers used in Pennsylvania, nutrients are dissolved completely in water. The major advantage is in handling. The disadvantages are the generally higher price and lower possible analysis compared to dry fertilizers, especially when the material contains potassium.

Suspension fertilizers, which are much less common, are fluids in which the components' solubility has been exceeded and in which very fine undissolved particles are kept from settling out by the inclusion of clay. Again, the major advantage of these materials is in handling. Suspensions also can be formulated at much higher analyses than can the clear solutions. Analyses similar to those for dry materials are possible. The major disadvantage of suspensions is that they require constant agitation, even in storage. Furthermore, suspension fertilizer cannot be used as a carrier for certain other chemicals.

The bottom line in comparing fluid fertilizers with dry fertilizers on the basis of amount of plant food, is that they are *equal* in agronomic effectiveness when each is used properly.

Remember, when making calculations of fluid fertilizers, that the analysis is given as a weight percentage, not on a volume or "per-gallon" basis. Most fluids weigh between 10 and 12 pounds per gallon. The following is an example to illustrate the calculations.

One gallon of the 10–34–0 liquid weighing 11.68 pounds per gallon contains: 11.68 lb Fertilizer/gallon ×.10 N = 1.17 lb N/gallon 11.68 lb Fertilizer/gallon ×.34 P_2O_5 = 3.97 lb P_2O_5/gallon

If 5 gallon per acre of this fertilizer is used as a corn starter fertilizer, this would apply the following nutrients: 5 gallon/A × 1.17 lb N/gallon = 5.85 lb N/A 5 gallon/A 3.97 lb P_2O_5 per gallon = 19.85 lb P_2O_5/A

It would take a little less than 9 gallons of this liquid to equal 100 pounds of a dry material with the same analysis.

For comparing fluids with dry fertilizers, divide the weight per gallon into 2,000 to get the number of gallons per ton.

In the above example, the calculation is: 2,000 ÷ 11.4 = 175 gallons per ton

UNDERSTANDING PHOSPHORUS FERTILIZERS

Several phosphate fertilizers can be used to meet the phosphorus (P) requirements of Minnesota's crops. This extension folder is designed to:

- Give a general description of the processes used in the manufacture of commercial phosphate fertilizers;
- Describe differences and similarities for the various materials.

MANUFACTURING COMMERCIAL PHOSPHATE FERTILIZER

Rock phosphate is the raw material used in the manufacture of most commercial phosphate fertilizers on the market. In the past, ground rock phosphate itself has been used as a source of P for acid soils. However, due to low availability of P in this native material, high transportation costs, and small crop responses, very little rock phosphate is currently used in agriculture.

The manufacture of most commercial phosphate fertilizers begins with the production of phosphoric acid. A generalized diagram showing the various steps used in the manufacture of various phosphate fertilizers. Phosphoric acid is produced by either a dry or wet process. In the dry process, rock phosphate is treated in an electric furnace. This treatment produces a very pure and more expensive phosphoric acid (frequently called white or furnace acid) used primarily in the food and chemical industry. Fertilizers that use white phosphoric acid as the P source are generally more expensive because of the costly treatment process.

The wet process involves treatment of the rock phosphate with acid producing phosphoric acid (also called green or black acid) and gypsum which is removed as a by-product. The impurities which give the acid its color have not been a problem in the production of dry fertilizers. Either treatment process (wet or dry) produces orthophosphoric acid— the phosphate form that is taken up by plants.

The phosphoric acid produced by either the wet or the dry process is frequently heated, driving off water and producing a superphosphoric acid. The phosphate concentration in superphosphoric acid usually varies from 72 to 76%. The P in this acid is present as both orthophosphate and polyphosphate. Polyphosphates consist of a series of orthophosphates that have been chemically joined together. Upon contact with soils, polyphosphates revert back to orthophosphates.

Ammonia can be added to the superphosphoric acid to create liquid or dry materials containing both nitrogen (N) and P. The liquid, 10-34-0, is the most common product. The 10-34-0 can be mixed with finely ground potash (0-0-62), water, and urea-ammonium nitrate solution (28-0-0) to form 7-21-7 and related grades. The P in these products is present in both the orthophosphate and polyphosphate form.

When ammonia is added to the phosphoric acid that has not been heated, monoammonium phosphate (11-52-0) or diammonium phosphate (18-46-0) is produced depending on the ratio of the mixture. The P present in these two fertilizers is present in the orthophosphate form.

The cost of converting rock phosphate to the individual phosphate fertilizers varies with the process used. More importantly, the processes used have no effect on the availability of P to plants.

PHOSPHATE FERTILIZER TERMINOLOGY

Because of the number of products on the market, the selection of a phosphate fertilizer can be confusing. An explanation of some terminology may help to avoid some of the confusion. Some important terms are:

Water-Soluble

Fertilizer samples analyzed by a control laboratory are first placed in water and the percentage of the total phosphate that dissolves is measured. This percentage is referred to as water-soluble phosphate.

Citrate-Soluble

The fertilizer material that is not dissolved in water is then placed in an ammonium citrate solution. The amount of P dissolved in this solution is measured and expressed as a percentage of the total in the fertilizer material. Phosphate measured with this analytical procedure is referred to as citrate-soluble.

Available

The sum of the water-soluble and citrate-soluble phosphates is considered to be the percentage that is available to plants and is the amount guaranteed on the fertilizer label. Usually, the citrate- soluble component is less than the water-soluble component.

Table . Percentages of water-soluble and available phosphate in several common fertilizer sources.

		P_2O_5		
P_2O_5 Source	N (%)	Total (%)	Available (%)	Water soluble* (%)
Superphosphate (OSP)	0	21	20	85
Concentrated Superphosphate (CSP)	0	45	45	85
Monoammonium Phosphate (MAP)	11	49	48	82
Diammonium Phosphate (DAP)	18	47	46	90
Ammonium Polyphosphate (APP)	10	34	34	100
Rock Phosphate	0	34	38	0

*Water-soluble data are a percent of the total P2O5
Source: *Ohio Agronomy Guide.* Ohio Cooperative Extension Service Bull.472.

*Water-soluble data are a percent of the total P2O5
Source: *Ohio Agronomy Guide.* Ohio Cooperative Extension Service Bull.472.

ORGANIC PHOSPHORUS SOURCES

Organic P fertilizers have been used for centuries as the P source for crops. Even with the advent of P fertilizer technology processes, organic P sources from animal manures—including composts—and sewage sludge are still very important. From a fertilizer/nutrient management perspective, the major differentiating factor is the availability of P. As with any of the fertilizer products, especially those with varying analysis, chemical analysis should be done on these products. Then an availability coefficient should be used to determine the available P as a portion of the reported total P.

Phosphorus from manure or sludge should be comparable to P from inorganic fertilizer. Therefore, if a producer has a P recommendation for 30 lbs/A of P_2O_5, applying approximately 65 lbs of 18-46-0 (DAP) or 6 tons of 11-6-9 (manure, 80% available P coefficient) should provide equivalent results.

The P contained in organic P sources is a combination of inorganic and organic P. Essentially, all of the inorganic P is in the orthophosphate form, which is the form taken up by growing plants. Diet fed to the animal has some control on this chemical make-up. Consider P feed supplements and the fact that many of these could be considered P fertilizers as well. Generally, 45-70% of manure-P is inorganic P. Organic P constitutes the remaining total P. Much of the organic P is easily decomposable in the soil, but factors such as temperature, soil moisture, and soil pH all have a bearing on the P mineralization rate. The final decomposition product is orthophosphate P compounds.

The combination of the organic/inorganic P ratios in the organic P sources and the soil environment affect the availability coefficient for organic P. Most animal manure research interpretations indicate that approximately 60-80% of the total P is available to crops in the first year. Due to the chemical composition of other organic P sources such as bone meal, lesser amounts of plant available P compared to total P are expected.

PHOSPHORUS SOURCE ISSUES

Some of the most frequently asked questions about phosphate fertilizer are discussed in the paragraphs that follow.

Should I Use Liquid or Dry? The utilization of P by plants is not affected by the liquid or dry property of the fertilizer. Plant nutrient use in both liquid and dry fertilizers is affected by such factors as method of application, crop and root growth characteristics, soil test levels, and climatic conditions. The amount of water in a fluid fertilizer is insignificant compared to the water already present in the soils. Therefore, P in liquid P sources is not more available than P in dry materials — even in a dry year. The selection of a liquid or dry P source should be based on adaptation to the farmer's operation and economics.

Is Orthophosphate Better Than Polyphosphate? To answer this question, it's important to understand the difference between these two forms of phosphorus.

The phosphorus in the phosphoric acid used to make most dry phosphate fertilizers as well as a few liquids is in the orthophosphate form.

If ordinary phosphoric acid is heated, water is removed and the orthophosphate ions combine to form a polyphosphate. This process does not convert 100% of the orthophosphate ions into the polyphosphate form.

Most polyphosphate fertilizers will have 40 to 60% of the phosphorus remaining in the orthophosphate form.

In the soil, polyphosphate ions readily convert to orthophosphate ions in the presence of soil water. This conversion is rapid and, with normal soil temperatures, can be complete in days or less. This conversion process is enhanced by an enzyme called pyrophosphatase, which is abundant in most soils.

Polyphosphates are usually marketed as liquid ammonium polyphosphate fertilizers. Because water is removed in the manufacturing process, these materials have a higher analysis than materials in which the phosphate is in the orthophosphate form. The polyphosphate liquids are also more convenient for the fertilizer dealer to handle and allow for the formulation of blends that are not possible with the orthophosphate liquids.

The effect of orthophosphate and polyphosphate fertilizers on crop production has been evaluated with numerous field trials.

Table .The influence of P source on corn yield.

P_2O_5 Applied	P Source	
	Polyphosphate	Orthophosphate
lb./acre	bu./acre	
15	124	124
30	134	134
45	142	142
Source: Nebraska Soil Test P: Low		

The averages from five sites where the soil pH was in excess of 7.3. It's obvious that the form of phosphate had no effect on yield and if there is a rapid conversion from polyphosphates to orthophosphates, these results are to be expected. Similar results from other studies have been reported throughout the Corn Belt.

Should Soil pH Influence Fertilizer P Source Selection? Soil pH should not be an important factor in selecting fertilizer P sources. From an academic perspective, monoammonium phosphates (MAP) create a more acidic zone around each fertilizer granule, whereas diammonium phosphates (DAP) create a basic zone.

Thus, in high pH soils, it can be theorized that using MAP-based fertilizers should be better than DAP because the acid-producing fertilizer would offset the calcareous soils. An additional concern regarding MAP or DAP selection, aside from soil pH, is potential ammonia toxicity to germinating seeds in dry soils. In applying the recommended amount of P in a drill-row or pop-up fertilizer placement, DAP will contain approximately 60% more N, which may be a potential injury risk. However, since agronomic studies and economic data indicate no crop yield differences, it can be concluded that fertilizer selection should be made on traditional factors such as nutrient content, price, availability, etc.

PHOSPHATE RICH ORGANIC MANURE AS FERTILIZER PHOSPHATE RICH ORGANIC MANURE AS FERTILIZER

Certain types of rock phosphates of sedimentary origin are applied to acidic soils as fertilizer directly. Recent research in India showed that high grade rock phosphate in fine size when used as a mix with organic manure works as efficiently as di-ammonium phosphate in alkaline soils. The rock phosphate from Jhamarkotra, India, Egypt and South Africa showed good agronomic efficiency when used as Phosphate Rich Organic Manure. Further PROM shows equal residual effect. The results of efficiency of PROM made using two grades (+ 24% P2O5 and 34% P2O5) of rock phosphate mineral from Jordan show that low grade rock phosphate slimes are as effective as high grade phosphate concentrate when used in PROM. Also PROM is very effective even in the saline soils where DAP completely fails.

The world today consumes around 150 million tones of high grade phosphate mineral, 90% of which goes into the production of chemical phosphatic fertilizers such as di-ammonium phosphate (DAP), single super phosphate (SSP) etc which contain P in water soluble form.

Phosphorous is an important nutrient element for plants. The element P is a constituent of DNA and RNA molecules as well as ADP and ATP, the molecules that transfer energy, facilitating biochemical processes in the plants.

Plants exude organic acids through their roots that dissolve phosphates naturally present in the soil, into their water soluble forms viz.

$$(H_2PO_4)^-, (HPO_4)^{-2}, (PO_4)^{-3}$$

which in turn are taken up by the plants. A variety of microorganism present in the soil organic matter release organic acids that can dissolve soil phosphates. Single Super Phosphate (SSP) that contains P in water soluble form was produced by the scientists of Rothamsted Experimental Station (England) in the year 1840 by reacting rock phosphate and sulfuric acid, with the idea of providing readily available P to the plants.

This is followed by production of more complex phosphatic fertilizers that contain water soluble P. Increased agricultural production world over is partly due to the introduction of chemical fertilizers that contain NPK. Unfortunately 60% to 70% of the P applied to the soils, in the water soluble forms is unavailable to the plant as applied phosphorous is fixed by Fe, Al, Mn ions in acidic soils and by Ca, Mg ions in alkaline soils into complexes that plants cannot take up.

Phosphorous availability to the plants is maximum in the narrow soil pH range between 5.5 and 7. Excessive and indiscriminate application of chemical fertilizers show adverse impact on the soils in that soil micro flora and fauna are destroyed thereby resulting into decreased agricultural production after years of application.

The phosphate rocks having high content of P soluble in 2% citric acid are considered for direct application in acidic soils. Phosphate rocks of sedimentary origin, that has PO43- partly replaced by CO32- isomorphically, show high content of P soluble in 2% citric acid. On the other hand phosphate rocks of igneous origin show less solubility of P in 2% citric acid and hence they are not considered for direct application in acidic soils.

Often organic manures are assessed in terms of their content of nutrient elements such as N, P, K, etc. Microorganism that decay organic matter produce a variety of useful compounds such as gibberlins, auxins, vitamins, fulvic and humic acids that are vital for the plant growth, therefore fertilizers and minerals cannot replace manure in agriculture. Peat or lignite can partly replace organic manures or composts for they also contain humic acids. Organic matter in the soil greatly enhances the water holding capacity of the soil in addition to facilitate the aeration by keeping the soil loose.

PHOSPHATE RICH ORGANIC MANURE

It is observed that farm yard manure (FYM) enriched with high grade (+34% P2O5) rock phosphate in fine size (d80 at 23 microns) shows better agronomic efficiency than di-ammonium phosphate when applied on equal P2O5 basis. Some initial results using Jhamarkotra rock phosphate (1T). Few companies in India are now producing and marketing PROM on commercial scale. The advantage with PROM is that it shows equal residual effect, that is it works for two consequent crops.

The dissolution of P from PROM is slow and is due to the organic acids released by the plant roots and microorganism hosted (naturally present or advertently added) by the soil and the soil organic matter. Further organic matter complexes soil cations thereby preventing fixation of P.

PHOSPHATE ROCK CHARACTERISTICS AND PROM

Each type of rock phosphate shows a characteristic curve of increasing content of P soluble in 2% citric acid (Y axis) as the particle size decreases (X axis). More interesting is the fact that citric acid (2%) soluble P2O5 content of even the rock phosphate of igneous origin increases substantially.

Studying these three types of phosphate minerals in PROM, Pareek et.al., report comparable yield of Vigna unguiculata (L) walp, to that of DAP on equal P2O5 basis.

A comparison of the performance of the rock phosphates of Jhamarkotra (India) High-Grade Ore (2T), Phalaborwa, SA Concentrate (3T), Egyptian High-Grade Ore (4T) as PROM with DAP showed the following order:

3T (1.125) > 2T (1.115) > DAP (1.05) H” 4T (1.002) > Control (0.465)

The seed output per plant in grams is shown in parenthesis. Control is without application of phosphate in any form.

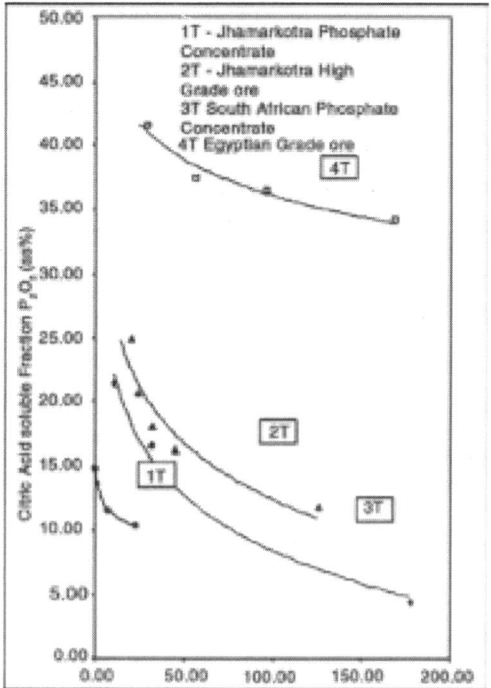

Fig. Increasing Fraction (as%) of P2O5 Soluble in 2%
Citric Acid of Rock Phosphate Mineral as the Particle size Decreases.

3

Mixed and Complex Fertilizers

CLASSIFICATION OF FERTILISERS

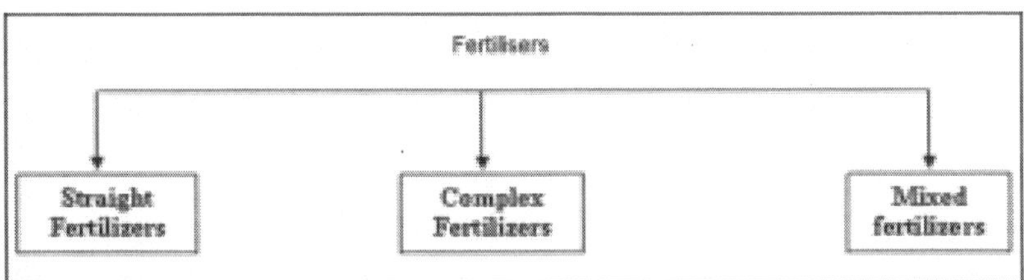

1. Straight fertilizers: Straight fertilizers are those which supply only one primary plant nutrient, namely nitrogen or phosphorus or potassium. eg. Urea, ammonium sulphate, potassium chloride and potassium sulphate.
2. Complex fertilizers: Complex fertilizers contain two or three primary plant nutrients of which two primary nutrients are in chemical combination. These fertilisers are usually produced in granular form. eg. Diammonium phosphate, nitrophosphates and ammonium phosphate.
3. Mixed fertilizers: are physical mixtures of straight fertilisers. They contain two or three primary plant nutrients. Mixed fertilisers are made by thoroughly mixing the ingredients either mechanically or manually. Fertilisers can also be classified based on physical form:

1. Solid 2. Liquid fertilizers

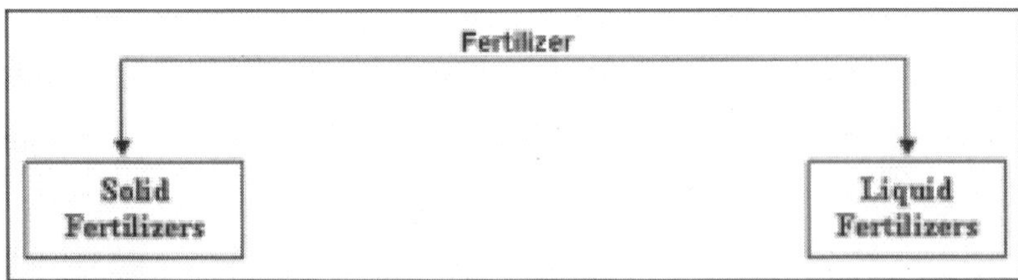

Solid fertilizers are in several forms *viz.*

1. Powder (single superphosphate),

2. Crystals (ammonium sulphate),
3. Prills (urea, diammonium phosphate, superphosphate),
4. Granules (Holland granules),
5. Supergranules (urea supergranules) and
6. Briquettes (urea briquettes).

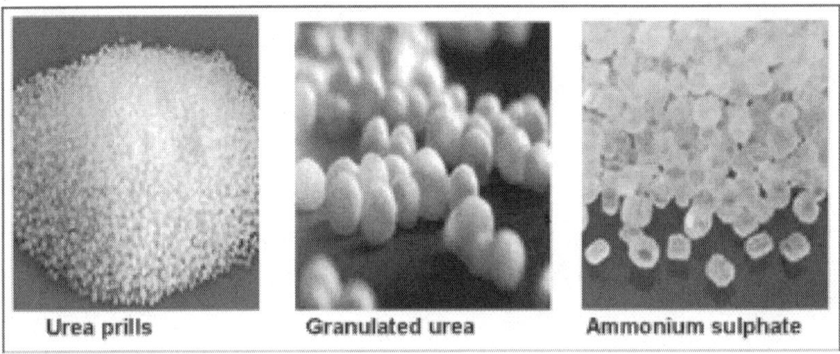

Liquid fertilizers:

1. Liquid form fertilizers are applied with irrigation water or for direct application.
2. Ease of handling, less labour requirement and possibility of mixing with herbicides have made the liquid fertilisers more acceptable to farmers.

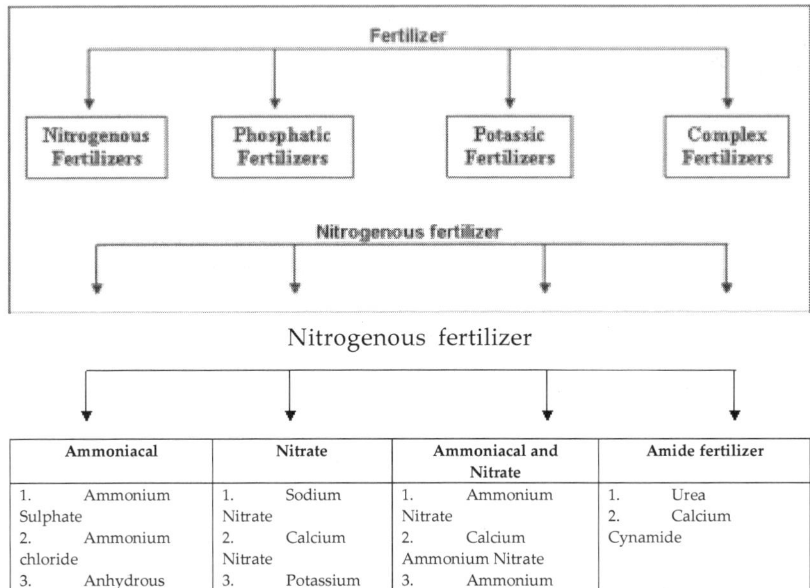

Nitrogenous fertilizer

Ammoniacal	Nitrate	Ammoniacal and Nitrate	Amide fertilizer
1. Ammonium Sulphate 2. Ammonium chloride 3. Anhydrous ammonia	1. Sodium Nitrate 2. Calcium Nitrate 3. Potassium Nitrate	1. Ammonium Nitrate 2. Calcium Ammonium Nitrate 3. Ammonium Sulphate Nitrate	1. Urea 2. Calcium Cynamide

NITROGENOUS FERTILIZERS

1. Nitrogenous fertilizers take the foremost place among fertilizers

since the deficiency of nitrogen in the soil is the foremost and crops respond to nitrogen better than to other nutrients.

2. More than 80 per cent of the fertilizers used in this country are made up of nitrogenous fertilizers, particularly urea.
3. It is extremely efficient in increasing the production of crops and the possibilities of its economic production are unlimited.

Ammoniacal fertilizers

1. Ammoniacal fertilizers contain the nutrient nitrogen in the form of ammonium or ammonia.
2. Ammoniacal fertilizers are readily soluble in water and therefore readily available to crops.
3. Except rice, all crops absorb nitrogen in nitrate form. These fertilizers are resistant to leaching loss, as the ammonium ions get readily absorbed on the colloidal complex of the soil.

a) *Ammonium sulphate* [$(NH_4)_2$ SO_4]

1. It is a white salt completely soluble in water containing 20.6.per cent of nitrogen and 24.0 per cent of sulphur.
2. It is used advantageously in rice and jute cultivation.
3. It is easy to handle and it stores well under dry conditions. But during rainy season, it sometimes forms lumps.
4. It can be applied before sowing, at the time of sowing or as a top-dressing to the growing crop.

b) *Ammonium chloride* (NH_4Cl)

1. It is a white salt contains 26.0 per cent of nitrogen.
2. It is usually not recommended for tomato, tobacco and such other crops as may be injured by chlorine.

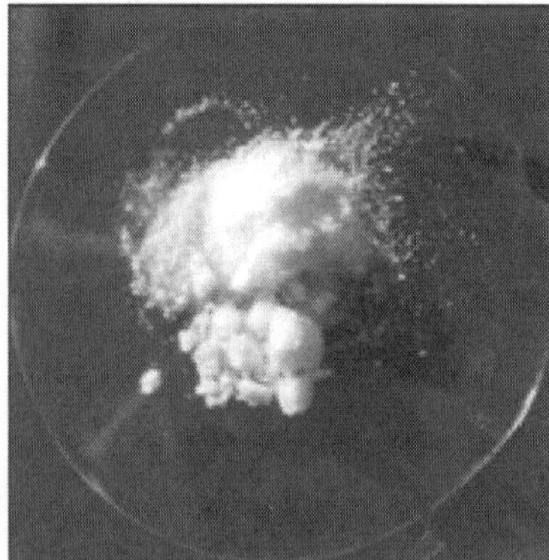

c) *Anhydrous ammonia* (NH_4)

1. It is a colourless and pungent gas containing 82.0 per cent nitrogen.
2. It is the cheapest and can be applied directly to soil by injection using blade type applicator having tubes.

3. It becomes liquid (anhydrous ammonia) under suitable conditions of temperature and pressure.

Nitrate Fertilizers

1. Nitrate fertilizers contain the nitrogen in the form of NO_3
2. These ions are easily lost by leaching because of the greater mobility of nitrate ions in the soil.
3. Continuous use of these fertilizers may reduce the soil acidity as these nitrogenous fertilizers are basic in their residual effect on soils.

a) *Sodium nitrate* ($NaNO_3$)

1. Sodium nitrate is a white salt containing about 15.6 per cent of nitrogen.
2. It is completely soluble in water and readily available for the use of plants as such, without any chemical change in the soil.
3. It is easily lost by leaching and denitrification.
4. When large quantities of sodium nitrate are added year after year, the nitrate ions are absorbed by crops and sodium ions accumulate and affect the structure of the soil. Sodium nitrate is also known as *chile salt peter* or *chilean nitrate.*
5. Sodium nitrate is particularly useful for acidic soils

b) *Calcium nitrate* [Ca $(NO_3)_2$]

1. It is a white crystalline hygroscopic solid soluble in water containing 15.5 per cent nitrogen and 19.5 per cent calcium.
2. The calcium is useful for maintaining a desirable soil pH.

Fig. *Calcium nitrate*

c) *Potassium nitrate* (KNO_3)

1. The purified salt contains 13.0 per cent nitrogen and 36.4 per cent potassium.
2. The nitrogen of the potassium nitrate has the same properties and value as that of the sodium nitrate.

Ammoniacal and nitrate fertilizers

These fertilizers contain nitrogen in both ammonium and nitrate forms. The nitrates are useful for rapid utilization by crops and the ammonical is gradually available.

a) *Ammonium nitrate* (NH_4NO_3)

1. It is white, water soluble and hygroscopic crystalline salt containing 35 per cent nitrogen half as nitrate nitrogen and half in the ammonium form.
2. In the ammonium form, it cannot be easily leached from the soil.
3. This fertilizer is quick-acting, but highly hygroscopic and not fit for storage.
4. It has an acidulating effect on the soil.
5. It is dangerous in pure form because of explosion hazard.

b) *Calcium ammonium nitrate* (CAN)

1. Calcium ammonium nitrate is a fine free-flowing, light brown or grey granular fertilizer, containing 26 per cent of nitrogen.
2. It is almost neutral and can be safely applied even to acid soils.
3. Half of its total nitrogen is in the ammoniacal form and half is in nitrate form.
4. It is made harmless by adding lime.

Fig. Calcium ammonium nitrate

c) *Ammonium sulphate nitrate* [$(NH_4)_2SO_4$ NH_4NO_3]

1. It contains 26 per cent nitrogen, three fourths of it in the ammoniacal form and the rest (6.5 per cent) as nitrate nitrogen.
2. In addition to nitrogen it contains 12.1percent sulphur.

3. It is a mixture of ammonium nitrate and ammonium sulphate.
4. It is available in a white crystalline form or as dirty-white granules.
5. It is readily soluble in water and is very quick-acting.
6. Its keeping quality is good and it is useful for all crops.
7. Its acid effect on the soils is only one-half of that of ammonium sulphate.
8. It can be applied before sowing, at sowing time or as a top-dressing.

Amide fertilizers

1. Amide fertilizers are readily soluble in water and easily decomposable in the soil.
2. The amide form of nitrogen is easily changed to ammoniacal and then to nitrate form in the soil.

a) *Urea* [CO $(NH_2)_2$]

1. It is the most concentrated solid nitrogenous fertilizer, containing 46 per cent nitrogen.
2. It is a white crystalline substance readily soluble in water.
3. It absorbs moisture from the atmosphere and has to be kept in moisture proof containers. It is readily converted to ammoniacal and nitrate forms in the soil.
4. The nitrogen in urea is readily fixed in the soil in an ammoniacal form and is not lost in drainage.
5. Urea sprays are readily absorbed by plants.
6. It may be applied at sowing or as, a top-dressing.
7. It is suitable for most crops and can be applied to all soils.

b) *Calcium cyanamide* ($CaCN_2$)

1. Calcium cyanamide or nitrolime contains 20.6 per cent of nitrogen.
2. It is a greyish white powdery material that decomposed in moist soil giving rise to ammonia.

PHOSPHATIC FERTILIZERS

1. Phosphatic fertilizers are chemical substances that contain the nutrient phosphorus in absorbable form (Phosphate anions) or that yield after conversion in the soil.

Super phosphate [Ca $(H_2PO_4)_2$)

1. This is the most important phosphatic fertilizer in use.
2. It contains 16 Per cent P_2O_5 in available form.
3. It is a grey ash like powder with good keeping or storage qualities.
4. Phosphatic fertilizer hardly moves in the soil and hence they are

placed in the, root zone.

Triple super phosphate:

1. The concentrated super phosphate is called as *Triple super phosphate* and it contains 46 per cent P_2O_5.
2. This fertilizer is suitable for all crops and all soils.
3. In acid soils, it should be used in conjunction with organic manure.
4. It can be applied before or at sowing or transplanting.

POTASSIC FERTILIZERS

1. Potassic fertilizers are chemical substances containing potassium in absorbed form (K+).
2. There are two potassium fertilizers *viz.*, muriate of potash (KCI) and sulphate of potash (K_2SO_4).
3. They are water soluble and so are readily available to plants.

a) *Potassium chloride* (KCI)

1. Potassium chloride or muriate of potash is a white or red, crystal containing 60.0 per cent K_2O.
2. It is completely soluble in water and therefore readily available to the crops.
3. It is not lost from the soil, as it is absorbed on the colloidal surfaces.
4. It can be applied at sowing or before or after sowing.
5. The chlorine content is about 47.0 per cent.
6. Its chlorine content is objectionable to some crops like tobacco, potato, etc where quality is the consideration.

b) Potassium sulphate (K_2SO_4)

1. Potassium sulphate or sulphate of potash is a white salt and contains *48* per cent K_2O.
2. It is soluble in water and therefore readily available to the crop.
3. It does not produce any acidity or alkalinity in the soil.
4. It is prefered for fertilization of crops like tobacco, potato etc., where quality is of prime importance.
5. It is costly because it is made by treating potassium chloride with magnesium sulphate..

SECONDARY MAJOR-NUTRIENT FERTILIZERS

Magnesium fertilizers

These are chemical substances containing the nutrient magnesium in the form of magnesium cations (Mg^{2+}).

Magnesium Sulphate ($MgSO_4$)

The utilization rate of magnesium fertilizers decreases w,ith incr,easing potassium supplies.

b. Calcium fertilizers

1. These are the chemical substances containing the nutrient calcium in absorbable calcium cations (‘Ca^{2+}) form.
2. The raw material of calcium fertilizers is lime found in nature.

Calcium Chloride ($CaCl_2$ $6H_2O$)

1. It contains at least 15 per cent calcium.
2. It is highly water soluble and can, therefore, be dissolved for application as a foliar nutrient.

c. Sulphate Fertilizers

1. These are chemical substances containing the nutrient sulphur in the form of absorbable sulphate anions (SO_4^{2-}).
2. The sulphur requirements of plants are about two third of their phosphorus requirements.
3. Substantial sulphur supplies occur as minor constituents of various N, P and K fertilizers.
4. Fertilization with sulphur becomes necessary with increasing removal from the soil with rising agricultural production especially in plants with high sulphur requirements. e.g. mustard

MICRONUTRIENT FERTILIZERS

1. The importance of fertilization of crops with micro-nutrients is increasing mainly because of greater removal from the soil, intensive liming of soil, intensive drainage of soil, higher use of nitrogenous, phosphatic and potassic fertilizers etc.
2. There are seven essential micronutrients required by plants.

These are iron, manganese, zinc, copper, chlorine, boron and molybdenum.

Iron fertilizers

1. These are generally water soluble substances, predominantly sprayed as foliar nutrients on the crops.
2. Plants absorb iron in the form of Fe^{2+}.

Commonly used iron fertilizers are as follows.

Ferrous sulphate **($FeSO_4$ $7H_2O$)**	It is a water soluble fertilizer containing 20 % Fe
Fe – Chelates Fe-EDTA Fe-EDDPA	Suitable for application as foliar nutrients

The manganese (Mn) fertilizers are as follows:

Manganous Sulphate ($MnSO_4 7H_2O$)	It is the well known water soluble Mn fertilizer. It is pink salt containing 24 % Mn. It dissolves in water and is suitable for foliar application.
Mn – chelates (Mn – DTA)	It contains 13 % Mn. It plays an important role in the crop fertilization.

Zinc fertilizers

Zinc (Zn) fertilizers play an important role in Zn deficient areas.

Zinc sulphate ($ZnSO_4 7H_2O$)	It is water soluble whitish salt containing 23 % Zn. It is applied as foliar nutrient. Its acidic action causes corrosion damage to plants
Zinc-oxide(ZnO)	It contains 70 % Zn. It is slightly soluble in water It is used as slow acting foliar nutrient

Copper Fertilizers

Copper fertilizers have been used to correct copper (Cu), deficiencies.
Copper sulphate ($CuSO4 5H2O$) - 25 % Cu
Copper sulphate ($CuSO4 H2O$) - 36 % Cu

Boron Fertilizers

Borax ($Na_2B_4O 10H_2O$)	It contains 11 % B It is water soluble white salt It can be applied as a soil dressing or foliar application
Boric acid (H_3BO_3)	It contains 18 % B It is a white crystalline powder It is applied as a foliar nutrient

Molybdenum Fertilizers

Sodium molybdate ($Na_2MoO^4{}_2H_2O$)	It contains 40 % Mo
Ammonium molybdate ($(NH_4)_6Mo_7O_244H_2O$)	It contains 54 % Mo

1. These are water soluble salts which contain Mo
2. They are suitable for soil application and foliar application as wel

Fertiliser Grade

1. Fertiliser grade refers to the guaranteed minimum percentage of nitrogen (N), phosphorus (P) and potash (K) contained in fertiliser material.
2. The numbers representing the grade are separated by hyphens and are always stated in the sequence of N, P, and

For example, label on the fertilizer bag with a grade 28-28-0 indicates that 100 kg of fertiliser material contains 28 kg of N, 28 kg of *P* and no potash.

1. Different grades of fertilisers are available in India.
Some of them are: 28-28-0
20-20-0
14-35-14
17-17-17
14-28-14 etc.

FERTILIZERRATIO

It refers to the ratio of the percentage of N, P_2O_5 and K_2O in the fertilizer mixture e.g., the fertilizer grade 12-6-6 has a fertilizer ratio of 2:1:1.

SUPPLIERS OF PLANT NUTRIENTS

These are straight fertilizers added to supply the plant nutrient mentioned in the grade.

CONDITIONERS

These are low grade organic materials like peat soil, paddy husk, groundnut hulls etc., which are added to fertilizer mixtures during their preparation for reducing hygroscopicity and to improve their physical condition.

Fig. Peat soil

Fig. Paddy husk

FILLER

A filler is a weight make material like sand, soil, coal powder etc, added to the fertilizer ingredients so as to produce a mixture of the desired grade.

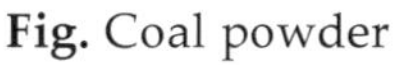

Fig. Coal powder

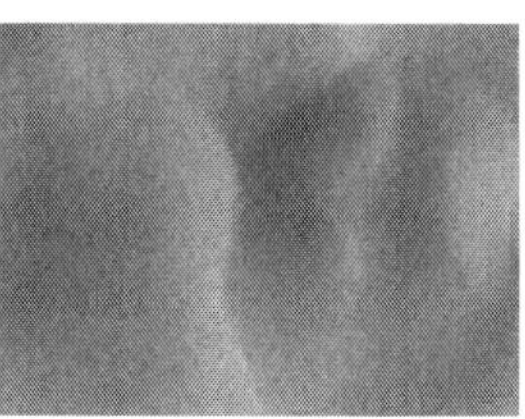

Fig. Sand

NEUTRALIZERS OF RESIDUAL ACIDITY

These are the materials like dolomite, lime stone etc, added in fertilizer mixtures to counteract the acidity of nitrogenous fertilizers.

FERTILIZER MIXTURES (FM)

When two or more fertilizers are mixed together to supply two or three major elements i.e. N, P_2O_5 and K_2O is known as fertilizer mixture or Mixed fertilizer. Or

A mixture of two more straight fertilizer materials is referred to as fertilizer mixture.Sometimes, complex utilizes containing two plant nutrients are also used in formulating fertilizer mixtures. Complete fertilizer refers to the fertilizers containing 3 major plant nutrients, N, P_2O_5 and K_2O.

Types of fertilizers: There are two types of fertilizer mixtures:

a. Open formula fertilizer mixtures: The formulae of such fertilizers in terms of kinds and quantity of the ingredients mixed are disclosed by the manufacturers.

b Closed formula fertilizer mixtures: The ingredients of straight fertilizers used in such mixtures are not disclosed by the manufacturers.

Materials used in fertilizer mixtures: Different materials go in to production of mixed fertilizers. In accordance with their principle function in the mixture, the materials can be grouped into:

1. Suppliers of plant materials: These are the straight fertilizers added to supply the plant nutrients mentioned in the grade, thus, are the primary materials most essential for preparing mixed fertilizers.
2. Conditioners: These are the organic substances which prepare the fertilizer mixture in good drilling condition and reduce caking. E.g.: Tobacco stems, Peat, Groundnut hulls and paddy hulls (Husks), bone meal, oilcakes.
3. Neutralizers of residual acidity: The substances used to neutralize the residual effects are known as neutralizers. For example, if the 'N'-ous fertilizers used are acididic in nature like Amm. Sulphate, Urea, a basic material like lime stone is added to counteract the acidity.
4. Filler: Filler is the make – weight material added to a fertilizer mixture. It is added to make up the differences between the weight of the added fertilizers required to supply the plant nutrients and the desired quantity of fertilizer mixture, such as sand, soil, ground coal ashes, sawdust and other waste products.
5. Secondary and micro – nutrients: Some times, secondary and micro – nutrient carrying fertilizers are added to correct its deficiency. An expression indicating the % of plant nutrient in a fertilizer

mixture is termed as fertilizer grade and the relative proportion of major plant nutrients in the mixed fertilizer taking 'N' as one, called as fertilizer ratio. For example, in a fertilizer mixture of 6:12:6 grades, the fertilizer ratio is 1:2:1.

The low analysis fertilizers contain less than 25% of primary nutrients and the high analysis fertilizers contain more than 25% of primary nutrients. On the other hand, the low analysis mixed fertilizers contain less than 14% sum of the primary nutrients and high analysis mixed fertilizers contain more than 14% sum of the primary nutrients.

Advantages:

1. The balanced fertilizer mixture suited to crop and soil can be supplied,
2. All the required nutrients can be supplied at one time by theapplication of fertilizers mixture and thus, time and labourers are saved.
3. Storage and handling costs are reduced.
4. Micro nutrients can be incorporated.
5. Mixtures have better physical condition and are easier for application.
6. Residual acidity can be neutralized by using neutralizers in mixture.

Disadvantages:

1. The cost of plant nutrients is higher than straight fertilizers.
2. All only one nutrient is required by the crop, the fertilizer mixtures are not useful and sometimes farmers may add nutrients in excess or in limited quantity.

COMMON MIXED FERTILIZERS

Fig. Mixed fertilizers provide essential nutrients to increase the health and size of garden plants.

Commercially available fertilizer mixes are designed to supply the correct ratios of nutrients for garden plants. The three main ingredients contained in mixed fertilizers are nitrogen, phosphorous and potassium. This is expressed on packaging as the N-P-K ratio. As with all garden chemicals, common mixed fertilizers should be stored out of reach of children. Once the fertilizer is diluted and applied, most treated plants are safe for children to play around. Check package labels for specific warnings.

WHAT IS A MIXED FERTILIZER?

There are three main types of fertilizers: straight, complex and mixed. Straight fertilizers supply one type of nutrient. Complex fertilizers are a single substance that supplies more than one nutrient. Mixed fertilizers combine more than one type of straight or complex fertilizer to give plants a specific mix of nutrients. Purchasing mixed fertilizers saves time and effort, since individual components of fertilizers do not have to be researched and mixed by hand. Commercially available mixes are balanced to provide the right amount of each nutrient, and often contain micronutrients that would be harder to mix at home.

ALL-PURPOSE FERTILIZERS

The main ingredients in all-purpose fertilizers for use on gardens and potted plants are chemical or natural sources of nitrogen, phosphorous and potassium. A water-soluble plant food with the ratio 24-8-16 contains 24 percent nitrogen, 8 percent phosphorous and 16 percent potassium. The remainder of the mix is filler or trace nutrients. One tablespoon of this fertilizer would be mixed with 1 gallon of water and applied every seven to 14 days when plants are actively growing. A time-release fertilizer to be worked into the soil every three months will have a different fertilizer analysis, such as 10-10-10.

LAWN FERTILIZERS

Mixed lawn fertilizers are typically sold in granular forms or as liquids that are applied when irrigating. For granular fertilizers, apply according to label directions, using a drop or cyclone spreader. Apply half the amount of fertilizer in one direction across the lawn, and then spread the remaining fertilizer in the other direction to avoid missing patches of lawn. Liquid fertilizers are often sold in a bottle that attaches to a garden hose and automatically dilutes the fertilizer as the lawn is watered. Lawn fertilizers contain high percentages of nitrogen and low levels of phosphorous. Typical ratios include 32-0-4 and 29-2-3.

SPECIAL PLANT FOODS

Different types of plants have different fertilizer requirements. A time-release food for shrubs might have a fertilizer ratio of 18-6-8 with traces of minerals like magnesium, copper and iron. Rose mixes contain higher amounts of phosphorous with similar proportions of nitrogen and potassium, such as a 9-18-9 or 18-24-16 ratio. Vegetables can usually be fertilized with all-purpose plant foods, although there are specific mixed fertilizers for a few plants. Tomato food, for example, has a fertilizer ratio of 18-18-21 and can be used on other vegetables as well.

STRAIGHT VS. MIXED FERTILISERS

Any commercial chemical which is added in the soil with intention to boost up the yields of the crop, is known as fertiliser. Based upon nutrient content, chemical form fertiliser are classified in different classes like simple, compound or mixed, slow released etc.

SIMPLE OR STRAIGHT FERTILISERS

Simple or straight fertilisers are designed to supply only one nutrient element. Depending upon the nutrient availability they are further classified as nitrogenous, phosphatic and potassic fertilisers. Ammonium nitrate in the form of 'Nitram', containing 34.5% of nitrogen, is a good example. Some simple fertilisers essentially used to supply one element may fortuitously provide another. For example ground mineral phosphate, essentially a phosphorus source, also contains some calcium. Similarly, ammonium sulphate, although designed to provide available nitrogen, contains rather more sulphur than nitrogen.

SIMPLE FERTILISERS

Nitrogen	**%N**
Urea	45
Ammonium sulphate	21
Prilled ammonium nitrate	34
Ammonium nitrate/calcium carbonate	21-26
Anhydrous ammonia	81
Liquid fertilisers containing ammonium nitrate, ammonia and urea	20-40
Phosphorus	**%P_2O_5**
Superphosphate	18-21
Triple superphosphate	45-47
Ground mineral phosphate	29-33
Basic slag	8-22
Potassium	**%k_2O**
Potassium chloride (muriate of potash)	60
Potassium sulphate	50

MIXED FERTILISER

A mixture of two or more straight fertiliser material is referred to as mixed fertiliser or fertiliser mixture. Mixed fertilisers are physical mixtures of fertiliser materials containing tow or three major plant nutrients. Mixed fertilisers are made by thoroughly mixing the ingredients either mechanically or manually.

Advantages

- Less labour is required to apply a mixture than to apply its various components separately. This is an important factor in areas where farm labour is scarce and expensive.
- If a proper mixture is used to suit a particular soil type and crop, the used of a fertiliser mixture leads to balanced manuring. This gives higher yield sand more profit to the cultivators.
- The residual acidity of fertilisers can be effectively controlled by the use of a proper quantity of liming materials in the mixtures.
- Micronutrients, which are applied in small amounts to soil can be incorporated in fertiliser mixtures. This facilitates uniform soil application of plant nutrients required in small quantities.
- Mixtures have a better physical condition (granulated) and are more easily applied than many straight fertilisers.

COMPLEX FERTILIZERS

Complex fertilizers or combined fertilizers - fertilizers that contain several nutrients (2-3 main nutrients). One granule simultaneously contains several plant nutritive elements. Complex fertilizers are characterized by concentrated combination of plant nutritive components. Complex fertilizers are produced as a result of interaction of initial components - ammonia, phosphoric and nitric acids etc. Complex-mixed fertilizers are obtained by mixing ready-made fertilizers and treating them with acids (sulfuric or nitric) and with ammonia. Mixed fertilizers - the product of mechanic mixing (as opposed to complex and complex-mixed fertilizers, where interaction occurs at a chemical level). When making mixed fertilizers generally superphosphate fertilizers with nitrogen and potassium chloride is being used.

The mixtures of simple fertilizers, out of which mixed fertilizers are being produced, are obtained in industrial environment or in blending equipments – this is called "dry" mixing. NPK 15:15:15 and NPK 16:16:16 refer to complex fertilizers – complex mineral fertilizers (nitrogen-phosphorus-potassium) in granules. NPK is used as the primary and pre-sowing fertilizer for all crops in all types of soils in Georgia. Complex fertilizers are attractive because of their universality; they are suitable for different needs of plants.

Georgian Agricultural production is continuously increasing and the usage of mineral fertilizers with increased concentration is becoming one of the ways of optimizing costs – decreasing the amount of fertilizers being transported, storage cost and application of fertilizers in soil.

4

Application of Manure Management in Fertilizer Sector

INTRODUCTION

The quantity, composition, and consistency of manure influence the selection and the design of manure-handling facilities. In its strictest definition, animal manure refers only to feces and urine. However, bedding, feed wastage, rain, soil, milk-house wastes or wash, and more are mixed with the feces and urine on many farms. This results in a manure that has properties considerably different from fresh manure. To minimize confusion, the term fresh will be added to manure when talking only about feces and urine mixtures.

NUTRIENT COMPOSITION OF MANURE AND COMPOSTNUTRIENT COMPOSITION OF MANURE AND COMPOST

Many different types of manure are available for crop production. It is assumed that most vegetable growers will be using solid manure with or without bedding. Similar principles will apply to the use of liquid manures. The nutrient content of manures varies with animal, bedding, storage, and processing. The approximate nutrient composition of various solid manures, including some composted manures.

While this table provides a general analysis of manure or compost nutrient content, it is strongly recommended that if routine applications are made for crop production the specific manure being used should be tested by a laboratory for moisture and nutrient content. Nutrient analysis should include: total nitrogen (N), ammonium-N, phosphate (P_2O_5), and potash (K_2O). Accurate manure or compost analysis requires that a representative sample be submitted; so several subsamples should be collected and composited to make up the sample. If manure or compost is being purchased, request a nutrient analysis from the seller for N, P_2O_5, and K_2O content.

Fresh *vs.* composted manure. Fresh, non-composted manure will generally have a higher N content than composted manure. However, the use of composted manure will contribute more to the organic matter content of the soil. Fresh manure is high in soluble forms of N, which can lead to salt build-up and leaching losses if over applied.

Fresh manure may contain high amounts of viable weed seeds, which can lead to weed problems. In addition, various pathogens such as *E. coli* may be present in fresh manure and can cause illness to individuals eating fresh produce unless proper precautions are taken. Apply and incorporate raw manure in fields where crops are intended for human consumption at least three months before the crop will be harvested. Allow four months between application and harvest of root and leaf crops that come in contact with the soil. Do not surface apply raw manure under orchard trees where fallen fruit will be harvested.

Heat generated during the composting process will kill most weed seeds and pathogens, provided temperatures are maintained at or above 131°F for 15 days or more (and the compost is turned so that all material is exposed to this temperature for a minimum of 3 days). The microbially mediated composting process will lower the amount of soluble N forms by stabilising the N in larger organic, humus-like compounds. A disadvantage of composting is that some of the ammonia-N will be lost as a gas. Compost alone also may not be able to supply adequate available nutrients, particularly N, during rapid growth phases of crops with high nutrient demands. Composted manure is usually more expensive than fresh or partially aged manure.

Heat-dried manure/compost. Drying manure or compost to low moisture content reduces their volume and weight, which lowers transportation costs, but it also requires energy inputs. Dried products can be easier to handle and apply uniformly to fields, especially those that have been processed into pellets. Heat drying also reduces pathogens if temperatures exceed 150 to 175°F for at least one hour and water content is reduced to 10 to 12 per cent or less.

The significant energy costs to heat-dry manure or compost at high temperatures are in contrast to the self-heating generated by microbial respiration during the composting process. Heat-dried composts vary widely in the degree to which they are composted before drying.

Many are only partially composted and have higher amounts of soluble (inorganic) N forms than mature, stable compost. This readily available N gives these products some characteristics that are similar to soluble N fertilizers, such as ammonium nitrate. Heat drying of manure and immature compost may increase volatilisation of ammonia-N and reduce the total N content of the finished product. In addition, composted or partially composted material that is dried at high temperature rather than going through a curing phase at ambient temperatures is not as biologically active as mature compost. The disease suppressive properties of some composts depends upon recolonisation of the compost by disease suppressing organisms during the curing phase.

MANURE AND COMPOST APPLICATION

Some of the N in fresh manure will be lost to the atmosphere during application in the form of ammonia gas. The higher the ammonium-N fraction is in manure, the more prone it is to ammonia volatilisation. Manure should be incorporated within 12 hours of application to avoid excessive ammonia losses.

Unincorporated manure will supply the organic N fraction and at most 20 per cent of the ammonium-N fraction. Incorporation of composted manure is not as critical, because the N is stabilised in organic compounds with little free ammonium present. However, in order to obtain full benefit from compost, incorporation is recommended whenever possible. Manure and compost are often high in soluble salts, so to avoid salt injury seeding operations should take place about 3 to 4 weeks after application.

RESIDUAL NUTRIENTS IN SOIL FROM MANURE AND COMPOST APPLICATION

The residual effects of the manure and compost are important. Some benefit will be obtained in the second and third years following application. When manure and compost are used to fertilize crops, soil organic matter will increase over time and subsequent rates of application can generally be reduced because of increased nutrient cycling. Continuous use of manure or compost can lead to high levels of residual N, P, and other nutrients, which can potentially be transported to lakes and streams in run-off or leach and pollute the groundwater. Taking into account residual release of N in subsequent years should help to avoid excessive applications. General rules of thumb for N are that organic N released during the second and third cropping years after initial application will be 50 per cent and 25 per cent, respectively, of that mineralised during the first cropping season.

Remember that some manures and composts contain high levels of P, so if organic nutrient sources are regularly applied at rates to meet crop N demands, the amount of P in the soil can build up to excessively high levels. Use of soil tests, plant tissue tests, and monitoring of crop growth will help in determining the amount of residual N and other nutrients in the soil and the need for further applications.

CALCULATING THE AMOUNT OF MANURE OR COMPOST TO APPLY

Methods for calculating the amount of manure or compost to apply have been adapted and summarised from *Livestock Waste Facilities Handbook*,

2nd ed., 1985, Midwest Plan Service. Composts can be thought of as similar to manure, but with little or no ammonium-N present. The amount of compost required to meet crop nutrient demands can be very large. For these situations, more readily available nutrients from other sources may be required to supplement compost additions, especially early in the growing season.

SAMPLING, TESTING, AND EVALUATING MANURE NUTRIENTS

SAMPLING

To make the most appropriate and environmentally responsible use of manure, it is necessary to test the manure to accurately determine its nutrient concentrations. Furthermore, for the test results to be meaningful, it is important to obtain an adequate number of samples that are representative of the bulk manure. The number of samples needed will depend on the variability of the manure. The more variable the manure, the more samples are required. Variability of liquid manures is usually less than for solid manures, especially if the liquid manure can be mixed prior to sampling.

Manure from different storage systems should be sampled differently. For solid manures, the sample can be taken while loading or during spreading. For poultry, the sample can come from within the house or from a stockpile of litter. Liquid manure can be sampled from storage or during the time of application. Liquid manure in storage should be agitated two to four hours before taking the sample.

Testing: Select a reputable testing laboratory. In selecting a laboratory, determine if the laboratory is knowledgeable in testing manure and belongs to a manure proficiency testing programme. Obtain the laboratorys appropriate sampling kit and read the corresponding instructions thoroughly. Manure samples should be identified regarding the farm, animal species, and date of sampling. The samples should be kept frozen until shipped to a laboratory. It is best to ship early in the week.

The laboratory analyses required on manure are moisture, total nitrogen (N), ammonia-N (NH_3-N), phosphorus (generally reported as phosphate [P_2O_5]), and potassium (generally reported as potash [K_2O]). Other useful analyses include pH, electrical conductivity, calcium (Ca), magnesium (Mg), and sulphur (S). The nutrients manganese (Mn), copper (Cu), and zinc (Zn) may also be important, especially if these nutrients are included in the animal diet.

EVALUATION OF TEST

Most laboratories design their test reports to meet the needs of their customers. However, there may be differences from laboratory to laboratory in the reporting format, reporting units of the test values, conversion factors, and the estimate of the nutrient availability. The test report usually involves three types of information. The first type is the descriptive information about the sample and the customer, including sample identification, sample description, and date of analysis.

The second type of information is the actual analytical results. Careful attention should be paid to the units associated with the test values. Typically, the laboratory will report the test values in parts per million (ppm) or per centage (per cent), and then convert them to the units needed by the customer to calculate application rates. Examples are lb per 1,000 gal for liquid manure and lb per ton for solid manure. Laboratories may also report the analysis on an as-is basis and a dry-matter basis. The as-is basis is used to calculate the application rates. Most laboratories report the results only on an as-is basis. However, the dry-matter basis, if it is needed, can be calculated from the moisture determination. The dry-matter basis is used to compare the nutrient concentrations of one manure with those of another manure.

The third type of information is interpretive and includes estimates of nutrient availability and fertilizer value. Since, the laboratory may be using estimates of nutrient availability from other sources, it is important to verify that the labs nutrient availability factors are the same as reported in this bulletin for Ohio conditions. The fertilizer value is often assigned to the manure by taking into account the current local fertilizer prices for the nutrients in the manure. However, the fertilizer value may be less when not all nutrients are needed due to already high soil-test levels. Manure application costs vary between fields and can be more than the fertilizer value of the nutrients.

APPLICATION OF MANUREAPPLICATION OF MANURE

Manure is a valuable resource that needs to be managed effectively and efficiently. Land application of manure should not be considered simply a disposal system. Manure provides nutrients for crops and helps build and maintain soil fertility. Manure can also improve soil tilth, increase water-holding capacity, lessen wind and water erosion, improve aeration, and promote beneficial organisms.

There are three principal objectives in applying animal manure to land:

- Ensure maximum utilisation of the nutrients in the manure by crops.
- Minimize environmental hazards.
- Minimize neighbourhood complaints and concerns.

Available land for manure application is an important consideration for all livestock operations. When planning a new operation or expanding an existing operation, adequate land area for manure application must be included in the plan. A conservative approach in determining the amount of land required is to consider the removal of the nutrient by the harvested crop.

This will ensure that enough land area is available in future years to prevent nutrient buildup in the soil beyond recommended agronomic and environmental levels. In addition to balancing nutrients, best management practices (BMPs) for applying manure to crops must be used. To maximize manure nutrients while minimizing potential environmental impacts and neighbours concerns, manure application must consider nutrient losses during handling and storage, run-off and preferential flow, and timing and rate of application.

APPLICATION OF MANURE ON FROZEN SOIL

Application on frozen and snow-covered soil is not recommended. However, if manure application becomes necessary on frozen or snow-covered soils, only limited quantities of manure should be applied. Frozen soil means that the soil surface is frozen so that manure cannot be injected into the soil profile.

If winter application becomes necessary, applications should be applied only if ALL the following criteria are met:

- Application rate is limited to 10 wet tons per acre for solid manure with more than 50 per cent moisture and five wet tons for manure with less than 50 per cent moisture. For liquid manure, the application rate is limited to 5,000 gallons per acre.
- Applications are to be made on land with at least 90 per cent surface residue (*e.g.*, good quality [grass] hay or pasture field, all corn grain residue remaining after harvest, all wheat residue cover remaining after harvest).
- Manure should not be applied on more than 20 contiguous acres. Contiguous areas for application are to be separated by a break of at least 200 feet. Utilize areas for winter manure application that are the farthest from streams, ditches, waterways, and surface water and use areas that present the least run-off potential.
- Increase the application setback distance to 200 feet minimum from all grassed waterways, surface drainage ditches, streams, surface inlets, and water bodies. Minimum suggested setback distances are doubled for winter application of manure to frozen or snow-covered soils. This distance may need to be further increased due to local conditions.
- Additional winter application criteria for fields with significant slopes of more than 6 per cent. Manure should be applied in alternating strips 60 to 200 feet wide generally on the contour, or in the case of contour strips on the alternating strips.

APPLICATION OF MANURE ON STEEP SLOPES

Manure should not to be applied to cropland with slopes of more than 15 per cent or to pastures/hayland with slopes of more than 20 per cent unless one of the following precautions is taken:

- Immediate incorporation or injection with operations done on the contour, UNLESS the field has 80 per cent ground cover (residue and/or canopy).
- Applications are timed during periods of lower run-off and/or rainfall.
- Lower rates can be applied by using split applications (separated by rainfall events). Apply no more than 10 wet tons/acre for solid manure, or 5,000 gallons/acre for liquid manure.
- The field is established and managed in contour strips with alternate strips in grass or legume.

Timing of manure application should also consider the potential impact on neighbours. To develop and maintain good neighbour relations, give adequate notice of the intent to land-apply manure and do not haul and spread on weekends, holidays, or important events. Good communication is key to minimizing neighbours complaints.

MANURE APPLICATION RECORD KEEPING

Keep good field records of soil and manure test results, yields achieved, and nutrients applied (time, form, rate, and method of application). Records should be kept for a period of five years or longer (metals analyses and associated application rates and locations should be maintained permanently):

- Quantity of manure produced and its appropriate analysis.
- The last three soil-test results.
- Dates, analysis, and amounts of manure that are land-applied.
- The dates and amounts of manure removed from the system due to feeding, energy production, or export from the operation.
- Manure application methods.
- Crops grown and yields (both yield goals and measured yield).
- Other tests, such as determining the nutrient content of the harvested product.
- Calibration of application equipment.
- Soil moisture and weather conditions (temperature and wind direction) at the time of application.
- Consider annual reviews to determine if changes in the nutrient budget are desirable (or needed) for the next planned crop.

FATALITIES

Fatalities may occur when people enter manure-storage structures, including covered manure pits, and are probably due to CO2 and H2S because these gases are heavier than air. Caution should also be taken when agitating manure as the asphyxiating effect of NH3, CO2, and CH4 combined with the toxic effect of H2S could be fatal.

Another potential risk, especially for children, is drowning in a pit, storage tank, and earthen storage basin or lagoon. Failure and breakage of slats or covers on pits and lack of protective barriers or railings around pit openings during agitation can lead to accidents. Push-off platforms or ramps (piers) can be a site for the tractor scraper and driver to tumble into an open storage structure or lagoon. Crusts on earthen storage basins can be a problem, especially for children, as they may appear capable of supporting ones weight, but they are not.

PRECAUTIONS

When designing manure structures and systems, think safety. When operating or managing manure equipment, think safety.

Consider the following major safety points when designing and operating manure equipment, structures, or systems:

- Do not enter a manure pit unless absolutely necessary and then only if:
  - — The pit is ventilated beforehand.
  - — You have supplied air to a mask or a self-contained breathing apparatus.
  - — You are wearing a safety harness and attached rope with at least two people standing by who are capable of pulling you out.
- When agitating a manure storage, always have at least one additional person available who can go for help if you are overcome by gases.
- Properly designed and operated ventilation systems can reduce the concentration of gases within the animal zone, improving animal performance. Poorly designed or improperly adjusted ventilation air inlets may actually increase gas concentrations at the animal level.
- When possible, construct lids for manure pits or tanks and keep access covers in place. If an open ground-level pit or tank is necessary, build a fence around it and post with Keep Out and Danger Manure Storage signs.
- Get help before attempting to rescue livestock or people that have fallen into a manure-storage structure.
- Build railings alongside all walkways or piers of open manure storage structures.

- Permanent ladders on the outside of above-ground tanks should have entry guards locked in place, or the ladder should be terminated above the reach of individuals.
- Never leave a ladder standing against an above-ground tank.
- Construct permanent ladders on the inside wall of all pits and tanks, even if covered. Use of non-corrosive material is important.
- Fence in earthen storage basins and lagoons and erect signs: Caution Manure Storage. The fence is also needed to keep livestock away from these structures. Additional precautions include a minimum of one lifesaving station equipped with a reaching pole and a ring buoy on a line.
- All push-off platforms or piers need a barrier strong enough to stop a slow-moving tractor. It should be low enough so that livestock cannot slide underneath.
- If possible, move animals before agitating manure stored in a pit underneath a building. Otherwise, if the building is mechanically ventilated, turn fans on full capacity when beginning to agitate, even in the winter, or if the building is naturally ventilated, do not agitate unless there is a brisk breeze. Watch animals closely during agitation, and turn off the pump at the first sign of trouble. The critical area of the building is where the pumped manure breaks the liquid surface in the pit.
- If manure storage is outside the livestock building, provide a water trap or other anti-back flow device to prevent storage gases from entering, especially during agitation.
- If an animal drops over, do not try to rescue it. You might become a victim of toxic gases. Turn off the pump, and do not enter the building until gases have had a chance to escape.
- Due to the possibility of explosion and fire, dont smoke, weld, or use an open flame in confined, poorly ventilated areas where methane can accumulate. Electric motors, fixtures, and wiring near manure-storage structures should be kept in good condition.
- Keep all guards and safety shields in place on pumps, around pump hoppers, on manure spreaders, tank wagons, power units, etc.

Take time now to review your total manure management system from a safety viewpoint. Think through each step of the collection system, storage or treatment units, and the land application phase. Are there dangerous areas in construction or operation? If so, make them safe. It could save your life or the life of a loved one or employee.

EMERGENCY ACTION PLAN

Every livestock farm should have an Emergency Action Plan in place. What is an Emergency Action Plan and why have one? It is a well-thought-out, simple, basic, common-sense plan that will help those involved with an emergency to make the right decisions.

A plan is needed:

- To meet the requirements of many states for a plan.
- To keep humans and livestock safe.
- To rectify an emergency situation.
- To protect the environment.
- To teach family members and employees.
- To record for future situations (prevention, law suits, etc.).
- To ensure notification of proper authorities.

SAFETY EQUIPMENT

Locate first-aid or rescue equipment near the manure-storage area. Clearly mark a wall closet or box and store the equipment inside it. Make occasional checks to ensure the equipment is in good order and has not been removed. Post the phone number of the local fire department/rescue squad on the wall beside the box and also near the telephone.

Personal protective equipment that includes air packs and face masks, nylon lines with snap buckles, and a parachute-type body harness with D rings for attaching lines can be obtained from supply sources of industrial safety and hygiene equipment. Look in the yellow pages under safety, safety equipment, industrial safety and hygiene, or safety supplies. These supply sources can also provide information on monitoring or measuring devices used to test hazardous atmospheres. Be sure to specify the gases you are dealing with when asking for or purchasing equipment.

Familiarize yourself, your workers, and your family with the proper operation of all safety equipment. Local medical (rescue) teams can assist in this education.

IMMEDIATE FIRST-AID PROCEDURES

- Do not attempt to rescue a victim from a hazardous gas situation unless you are protected with a supplied air-breathing apparatus.
- Have someone telephone for an emergency medical (rescue) squad, informing them there is a victim of toxic (manure) gas asphyxiation.
- If the victim is free from the immediate area of danger and there is no personal threat to your life, take the following steps:
  - — With the victim on his or her back, check for breathing, then give four quick mouth-to-mouth breaths and check for a pulse.
  - — If there is a pulse, continue mouth-to-mouth breathing every five seconds (12 per minute).
  - — If there is no pulse, start CPR (cardiopulmonary resuscitation) immediately. When the emergency squad arrives, the victim

should receive a high concentration of oxygen at the scene and in transport.

If members of your family have not taken CPR and first-aid training, enroll them in a course at your earliest opportunity. Periodic refresher courses in CPR are recommended.

MANURE MANAGEMENT AND EFFECTS OF MANURE ON THE ENVIRONMENT

MANURE MANAGEMENT SYSTEMS

Manure management systems are highly diverse, among which the following can be distinguished:

Grazing. Substantial losses through leaching may occur due to the uneven distribution of faeces and urine (urine patches may have a N load equal to 200-550 kg/ha; Van der Meer and Meeuwissen, 1989; Romney et al., 1994). Volatilization of N may also be considerable, but less important since part is deposited on nearby areas, though some of it on non-agricultural land.

Kraals. These enclosures are often used as in-situ fertilization of arable land by moving the kraal regularly. Soil fertility of a larger area, used for grazing, is partially concentrated on the arable land, thus enabling crop production in resource-poor situations. Losses through leaching will be slightly higher than during grazing as equivalent N and K fertilization rates are increased.

Dry lot storage. If urine is not collected and bedding is sparsely used, losses of N and K in particular will be high as most urine is lost. Depending on the storage facilities and storage time of the faeces part of the nutrients in faeces will also be lost through leaching and surface runoff, in the case of a precipitation surplus and uncovered manure heaps. Urine collection will minimize K losses but N losses will often remain high as volatilization will increase, though this is dependent on climatic conditions, storage time and storage method. Using bedding, with sufficient absorption capacity to capture urine, might reduce N losses with ca. 15% of the mineral N.

Slurry storage. This system of manure storage, where faeces and urine are stored together, is the main system in intensive livestock systems in OECD countries, except for broilers. Volatilization losses are dependent on the level of ventilation, depth of storage tanks and storage time, but often range between 5 and 35% of the total N excreted.

Lagoons. Lagoon systems are quite common at large livestock farms in Eastern European countries and, to a lesser extent, in Asia, while their importance is growing in the USA. Liquid manure, either before on after separating part of the solids, is treated in anaerobic lagoons. Organic material is decomposed, thereby mineralizing part of the nutrients. The liquid phase is either discharged into surface water or used for irrigation. The main problems are related to the discharge into surface water, leaching through the lagoon bottom, and odour. High NH_3 emission will occur as a major part of the N in mineral form, while also high CH_4 and N_2O emissions are also common.

Plastering material for house construction. This is particularly important in Africa, however the amount of manure involved on a global scale is considered to be too insignificant. In this system all nutrients are lost for agriculture.

Fuel. In many developing countries, and particularly in India, manure is an appreciable fuel. If burnt directly, most of the C, and all the N and S will be lost; other nutrients may be recycled to arable land via the application of the ash. The production of biogas from manure is another method to valorize the energetic value of manure. The high water content of the slurry makes it more difficult to handle, and N losses via volatilization may be high, because most N in slurry is in mineral form. Though strongly promoted (e.g. in China) and applied to some extent in Asia, its present application is still limited mainly due to high investment costs (both for the digester and adjoining equipment) and technical problems.

Feed. Manure could be recycled by feeding it to animals, both livestock and fish, but this practice is limited. In addition to widespread reluctance to use manure as feed, probably originating from fear of health hazards, this can be explained by the low nutritive value of most types of manure, except for poultry manure as ruminant feed which is of a reasonable quality (ca. 55-60% TDN, 20-30% CP). Consequently, in intensive production systems where collectable manure is abundant, more economic feed is available, while in production systems where the use of low quality feeds is common, high collection costs and/or opportunity costs (manure as fertilizer or fuel) are prohibiting the use of manure as feed. Therefore, no more attention will be paid to this subject in this report. A recent overview on this subject, however, is given by Sanchez (1994).

EMISSIONS BEFORE MANURE APPLICATION

Stables and manure storage are major sources of ammonia (NH_3) emission. About half of the N in manure (i.e. liquid manure or slurry) is NH_3-N in solution. Because of the high vapour pressure of the NH_3, it will readily volatilize upon exposure of the manure to the air. The greater the exposure, i.e. a larger specific area in contact with the air, the more NH_3 volatilization.

If manure is stored in direct contact with the soil, the liquid can seep into the soil and into the ground water. The nutrients N, P, K, organic and other compounds can leach into the ground water and the manure storage thus acts as a major source of ground water pollution.

Surface runoff from farmyards, kraals, bomas and manure storage can be an important source of pollution.

EFFECTS ON SOIL QUALITY

Manure application to agricultural land involves the addition of all the components of the manure to the soil. An appropriate balance should be maintained between agronomic requirements and negative environmental impacts. Negative impacts, that could be defined as soil pollution, have to do with the addition of heavy metals, organo-chlorines and too many salts. Also, weed seeds could be spread through manuring the land. On the other hand, manuring almost always has a positive influence on the build up of soil organic matter and thus improves the "intrinsic" fertility of the soil, as well as the soil structure.

After application of manure, decomposition by microorganisms of the organic material will start into carbon dioxide (CO_2), water (H_2O) and minerals of plant nutrients such as N, P, S and metals. The transformation of organically bound elements into plant available nutrients during microbiological decomposition is called mineralization. Organic matter that remains one year after application is assumed to be part of the soil organic matter and will decompose gradually over the years, releasing plant nutrients in a way that resembles a slow release fertilizer. A more fertile soil, consequently, has less need for mineral fertilizers. The fertilizer industry uses non-renewable resources such as fossil fuel, phosphorus and potassium deposits, and is a source of emissions. Manuring has, in an indirect way, a positive effect on the environment. A small fraction of the added organic material is transformed into organic matter that is resistant to microbiological breakdown, the so-called humus or stable organic matter. Humus contributes to soil fertility by retaining plant nutrients through adsorption. It also acts as binding material in the soil, improving soil structure and is responsible for making clay soil less susceptible to compaction caused by heavy traffic, or a silty soil less susceptible to erosion. In addition, humus increases the water holding capacity and the cation exchange capacity (CEC) of any type of soil.

The heavy metals Cu and Zn have been mentioned as major contaminants from the heavy application of pig slurry (e.g. in part of The Netherlands). Repeated application of large doses of pig slurry to the same plot may lead to Cu and Zn levels in the soil that are toxic, for instance, to soil fauna and sheep. Since the 1978 EC legislation Cu additions to pig feed have been reduced increasingly, to a level of 35 mg per kg for growing and finishing pig feeds. At current levels, Cu and Zn are considered not problematic if P fertilization is in balance with the crop requirements.

There is a danger of incomplete degradation of organo-chlorines by microorganisms. Through manuring, they could be taken up by crops and pose a threat to humans by accumulating somewhere in the food chain. Many countries have replaced organo-chlorines with organo-phosphates, but residues from insecticides still continue to be the main source of organo-chlorines in feed. Organo-chlorines originating from substances used against ecto-parasites can also be found.

The passage of weed seeds through the digestive tract of animals reduces their germination capability. Some weed species, however, survive. In a stack of farmyard manure, the temperature rises above 55 °C because of the microbiological decomposition of the organic matter and kills weed seeds within three weeks. The germination capability of weed seeds stored in slurry is destroyed only after a period of five months.

Manure contains much dissolved potassium chloride (KCl) and sodium chloride (NaCl). Repeated application of large amounts of manure in arid or semi-arid climates may easily lead to salinisation of the soil, making it unsuitable for many crops.

EFFECTS ON GROUND WATER AND SURFACE WATER QUALITY

The main dangers of the application of manure are runoff of manure or manure components into surface water and leaching of nitrate (NO_3) and P into the ground water. Mineral N in manure is largely present as NH_3. If, upon application of the manure, it does not volatilize, it will be quickly nitrified, i.e. transformed through microbial action into NO_3. Also, N mineralized from the organic fraction of the manure, will readily be nitrified. As NO_3 is an anion that is not adsorbed by clay minerals or soil organic matter, it is easily leached in case of a precipitation surplus. This holds good for NO_3-N from manure, and for that originating from mineral fertilizers or from decomposed soil organic matter. If ground water concentrations of NO_3 become too high, it is unsuitable for drinking water. Under certain conditions, ground water can flow into surface water. In brackish and salt water in particular high NO_3 concentrations in surface water will lead to eutrophication. Under certain conditions this may lead to excessive growth of algae, causing oxygen shortage and consequently the death of fish. Phosphorus is not nearly as mobile in the soil as NO_3 and therefore much less susceptible to leaching. Nevertheless, leaching of P can occur under certain conditions. Many sandy soils in The Netherlands have become "saturated" with phosphate (P_2O_5) after many years of heavy doses of manure. When saturated, the soil loses part of its capacity for P retention and leaching occurs. If P flows into the ground water and subsequently into surface water, the same problems as described above for NO_3 will occur. Note, P causes eutrophication in freshwater bodies in particular.

EFFECTS ON AIR QUALITY

Two processes involving N and one involving carbon (C) from manuring have an important effect on air quality. First, surface application of manure, particularly liquid manure, may cause substantial losses of NH_3 by volatilization. In the Netherlands, for instance NH_3 volatilization from manure is a major contributor to acid deposition. Unlike SO_2, a contributor to acid deposition from cars and industry, most of the emitted NH_3 is deposited near the emission source. Forests near regions with a high livestock density, are in a poor condition because of soil acidification caused by NH_3 deposition originating from the livestock industry. Acidification may lead to mobilization of aluminium (Al) ions, which are very toxic to fish, disturbs the nutrient uptake of plants and trees, and enhances sensitivity to stress factors like drought and fungi. Besides the acidifying effect, NH_3 deposition accounts for a considerable N load to the environment, causing eutrophication problems and N enrichment of the soils in nature reserves. The last mentioned can cause undesirable changes in species composition (important for biodiversity).

Second, denitrification of NO_3 by microorganisms is possible under anaerobic conditions when N_2 is formed, but giving off a by-product N_2O, a gas that affects the ozone layer. Although quantitative data are scarce, animal excreta and arable land may be important sources of N_2O globally.

A third important air pollutant is methane (CH_4), formed upon decomposition of manure under anaerobic conditions. If stored manure is disturbed, CH_4 will escape into the atmosphere and eventually, like N_2O, affect the ozone layer.

In addition to CH_4 formation in manure storage, the use of manure in flooded rice production (anaerobic conditions) and CH_4 formation in the rumen of ruminants are important sources of CH_4 emission.

Odour has a negative effect on the air quality, affecting animals and people in closed stables as well as people near farms producing or applying manure. Especially in combination with dust odour may cause serious health problems. In the USA it is estimated that 70% of the workers in closed animal stables suffer from respiratory illness, this is one of the main causes of labour disability of pig farmers in the Netherlands. Hydrogen sulphide and ammonia are the main contributors to odour problems, though a dozen other compounds, like fatty or organic acids, phenols, etc. may also play a role. Most of the odour comes from the anaerobic decomposition of manure.

EFFECTS ON CROPS

Manure is applied to agricultural land chiefly because of its fertilizing value. Animal manure supplies all major nutrients (N, P, K, Ca, Mg, S,) necessary for plant growth, as well as micronutrients (trace elements), hence it acts as a mixed fertilizer The fertilizing effect on crops can be compared to the effect of mineral fertilizers, and expressed in working coefficients. If, for example, the N in pig slurry on maize is half as effective in terms of yield increase as the N from ammonium nitrate (which is the reference chemical fertilizer), the working coefficient is 0.5. Manure application in a given year will influence not only crops grown that year, but also crops in subsequent years, because decomposition of the organic matter is not completed within one year. Working coefficients for subsequent years could be determined as well. Therefore, the application of manure, thus, saves mineral fertilizers for various nutrients. This illustrates that nutrients from animal manure can be substituted for mineral fertilizers and which is far better for the environment.

A disadvantageous aspect of the uptake of components from manure by the crop is over-dosage, which can lead to the absorption by plants of non-degradable components such as heavy metals (Cu, Zn) and organo-chlorines. These components can accumulate in the food chain and become a health hazard.

PARAMETERS INFLUENCING EFFECTS OF MANURE ON THE ENVIRONMENT

The effects of manure on the environment are strongly influenced by the prevailing climatic, soil and hydrological conditions. The rough estimates are made of the extent to which these conditions influence processes such as emission, leaching and crop utilization of nutrients from manure.

CLIMATE

The most important climatic characteristics affecting processes such as emission, leaching and decomposition of organic material are temperature, precipitation and evapotranspiration. Temperature strongly influences all microbiological processes. Higher temperatures, therefore, lead to higher rates of nitrification, denitrification and decomposition of organic material, but also to faster crop growth and the associated uptake of nutrients from manure. Nitrates are formed faster and are therefore more susceptible to leaching, but are also taken up faster by plants if a crop or vegetation is present. At higher temperatures and under reduced conditions NO_3 will, be denitrified more rapidly and N_2O, (harmful to the ozone layer), will be formed more quickly.

The extent of nutrients leaching to the ground water is largely determined by the balance of precipitation and evapotranspiration. If, for a certain period precipitation exceeds evapotranspiration considerably, NO_3, P, other nutrients and organo-chlorines can leach to the ground water. Conversely, if annual potential evapotranspiration greatly exceeds annual precipitation, upward movement of water in the soil will occur, with the imminent danger of salinisation. High doses of manure could add such large amounts of nutrients that they cannot all be taken up by the crop. Nutrients then accumulate and cause salinisation. Potential evapotranspiration (PET) also strongly influences the rate of NH_3 volatilization; high PET leading to high rates of volatilization.

SOIL

Of the soil characteristics, only texture and pH. The water and nutrient holding capacity of clay soils is higher than that of sandy and silty soils, therefore leaching of NO_3, P, other nutrients and organo-chlorines is dependent on soil texture. Conversely the risk of accumulation of harmful components in the root zone following repeated application of large doses of manure is higher in heavier textured soils than in light soils. Clay soils become more easily waterlogged after heavy rainfall because of a lower hydraulic conductivity, i.e. the possible rate of water transport through the soil. Under waterlogged conditions, denitrification can occur and harmful N_2O may be formed. Under extreme acid or alkaline conditions ($pH<4$ or>9), soils tend to deflocculate, the structure is destroyed and leaching of many organic and inorganic components becomes inevitable. Volatilization of NH_3 from soils with higher pH values is greater than from those with lower pH values.

HYDROLOGICAL CONDITIONS

The hydrological conditions strongly influence the leaching process, for example, faults or sinks in karst zones can cause leaching into deep, ground water-carrying layers. Pollutant containing topsoil can come into contact with deep aquifers via old ground water wells.

The flow from ground water to surface water is usually direct in areas with a shallow ground water level. In this way, leached nutrients/pollutants can flow rapidly from the soil via the ground water to the surface water and cause eutrophication and toxicity problems.

ENVIRONMENTAL IMPACT OF MANURE

BALANCING NUTRIENT IMPORT AND EXPORT

The effect on the environment of the manure produced in a particular agricultural system should be assessed by considering its role in the total nutrient management of the system. If the import and export of nutrients in the system is in balance, and animal manure is to play a positive role, it implies that losses from animal manure must be minimal. It also implies that efficient use is made of the manure in crop production, i.e. a large fraction of the nutrients from the manure is taken up by the crop. If, for example, NH_3 volatilization from the stables is high, a large proportion of the N imported into the system through the feed is lost. This is an indication of unbalanced N management of the farming system. During storage of manure in the open, NH_3volatilization and NO_3, P and K leaching occurs particularly when rainfall is high. This, again means nutrient losses. Surface spreading of manure without working in during periods of precipitation surplus on fallow land, may lead to NH_3 volatilization, NO_3 and P leaching, and surface runoff of manure. Eventually, the agricultural value of animal manure will appear to be low. Thus, the effects of manure management decisions early in the nutrient cycle will have a bearing on the later phases.

Van der Meer (1991) gives an example of an unbalanced system in the Netherlands. Of an intensive dairy farm,where the annual imports of N into the system largely exceed the exports. The total import of N per hectare is 533 kg through mineral fertilizers, imported feed and atmospheric deposition, while removal of N amounts to only 84 kg through milk and cattle weight gain.Thus, the annual N surplus of the system is, 449 kg N per hectare.

The extreme opposite of this system is presented in Sub-Saharan Africa, where Van der Pol (1992) estimates the annual nutrient depletion per hectare in a farming system in Southern Mali at 25 kg N, 20 kg K and 5 kg Mg. Stoorvogel *et al.* (1993) indicate an annual average nutrient loss per hectare from agricultural land in sub-Saharan Africa of 22 kg N, 2.5 kg P and 15 kg K for 1982-84, and estimate these values to be 26, 3 and 19 kg respectively in 2000. Here farmers derive a large part of their income from soil nutrient depletion or soil mining.

Animal densities in the Netherlands and Southern Mali differ considerably. An average Dutch dairy farm supports two to three highly productive livestock units (LU) per ha.

The overall animal density in the Netherlands, including intensive pig and poultry production and arable land, is over 4 LU per ha.

For Southern Mali, Van der Pol (1992) gives an overall animal density of 1.2 tropical livestock units (TLU) or 0.5 LU per ha of agricultural land.

Studying, the surplus and deficit situations in the above examples it is obvious that, manure management should aim at reducing the negative effects (lower nutrient losses) and maximizing the positive effects (plant nutrient supply and organic matter supply to the soil) of manure. A more balanced nutrient management will result with less burden on the environment.

Before an attempt can be made to assess the manure management in an agricultural system, at least the following key indicators should be explained:

- manure production, expressed as manure N or manure P production;
- NH_3 emission, expressed as fraction volatilized N from mineral N in the manure;
  surface water pollution by manure, expressed as biological oxygen demand (BOD) and N and P load per kg manure discharged;
- NO_3 leaching, expressed as fraction N leached from the soil of total N in manure;
- P leaching, for which the P balance per hectare will be used as an

indicator;

- agricultural value of the manure, expressed as fertilizer equivalents.

Some methods or rules of thumb for quantification of these indicators from literature data. The aim is to enable the livestock system specialists of this study to assess the impact of the manure produced in their respective systems.

MANURE PRODUCTION

MANURE PRODUCTION PER ANIMAL

The amount of manure produced by animals is very variable, even within species, partly due to differences in dry matter content of the manure. As water is not a very interesting component, we will be dealing mainly with the dry matter (DM), the organic matter (OM) and the nutrients N and P.

The little data on DM excretion reported in literature is difficult to interpret, partly because the amount of bedding included often varies, while the weight of the animals and the feeding situation is often poorly described. Müller (1980) gives manure production specifications for different animal species in different regions of the world but the basis for these estimates is unclear.

Ruminants	30-55
Pigs	10-35
Poultry	10-35

On the other hand, faeces production (without bedding) is measured in every *in vivo* digestibility trial, as the DM in faeces equal the indigestible component of the diet. Thus, if feed intake and apparent digestibility are known, faeces DM production can be calculated.

Although the prediction of ad lib feed intake is complicated and in practice animals are often not allowed to eat up to their maximum capacity, reasonable guesstimates can be made for most practical situations. Some indications for faeces DM production per species. Values are given in g per $kg^{0.75}$, as feed intake and consequently manure production is more close by related to metabolic weight to bodyweight. Lowest values are most likely to be associated with rations of very low and high nutritive quality (ruminants) or with rations of very high quality only (monogastrics with restricted feeding), though manure production is also influenced by palatability of the feed, genetic differences in feed intake capacity, and feed management. More accurate estimates can be made for specific situations if feed intake and digestibility are reasonably well known.

The values quantity of dry matter excreted by animals and not to the collectable amount. During storage, decomposition of organic matter starts, thereby reducing the quantity of collectable manure. The rate of decomposition is dependent on both environmental conditions and manure characteristics. Manure with a high content of easily degradable organic matter compounds, particularly poultry manure, may have high decomposition rates: up to 10% after one week and 40% after 3 months.

NUTRIENT EXCRETION PER ANIMAL

The positive and negative effects of nutrients in manure production, are more important than the dry matter. The nutrients excreted by animals can be derived by subtracting the nutrients converted into livestock (nutrient retention) products from the intake of nutrients via the feed. The typical nutrient retention values for major animal products, where variation especially in nutrient content per kg of live weight can be seen due to differences in the fat content of animals (fat animals having slightly lower N content per kg of live weight).

	N	P
Ruminants	26.0	7.5
Pigs	22.5	5.0
Poultry	28.0	4.0
Eggs	19.2	2.1
Milk (3:4% protein)	5.4	0.9

Average nutrient content of almost every feedstuff can be derived from standard feed composition tables. Estimation of feed intake and feed composition is a bigger problem, as the intake is unknown with the exception of some highly advanced year round stall-fed livestock systems However, some reasonable assumptions can be made for practical situations. Some examples for nutrient excretion of animals. Owing to the variation in intake and nutrient content of the feeds, these values represent examples and not averages for low or high productive situations.

NUTRIENT DISTRIBUTION BETWEEN FAECES AND URINE

A distinction between nutrients in faeces and in urine is often highly relevant as management for both types of manure output is often different: collection and utilization rates of urine are often much lower. To assess the quantity of NH_3 emitted and the agricultural value of manure a distinction between N in mineral form (N_m) and N in organic form is highly relevant. Though some N_m may be present in manure, mainly originating from manure decomposition, its concentration is generally negligible. Thus, N_m is in most practical situations equal to the quantity of N in urine.

Digestible protein values, often reported in older literature, are not very accurate for formulating feeding rations, as they do not represent the value of a feed for a particular animal. However, for the estimation of N in faeces, information on digestibility is very useful as the difference between crude protein (CP) intake and digestible crude protein (DCP) intake is the amount of N in the faeces, using the assumption that protein consists for

16% of N. The amount of N in urine can then be estimated as the difference between digestible N intake and N retention in livestock products.

N excreted via urine is prevalent in high production situations. This is particularly true if rations are unbalanced in terms of rumen degradable energy and rumen degradable protein. In low production situations, N excreted via urine is relatively lower, especially if feed rations are low in N (which is common for many animals in developing countries): e.g. urine from cows on sub-maintenance rations of straw will contain hardly any N as almost all N is recycled to the rumen instead of being excreted via the urine. In low production conditions, N in the urine is also lower, while N in faeces is higher, due to higher amounts of tannin in the feed which reduce rumen degradability of protein.

	Intake		**Retention**		**Excretion**	
Animal	N	P	N	P	N	P
Dairy cow (1)	163.7	22.6	34.1	5.9	129.6	16.7
Dairy cow (2)	39.1	6.7	3.2	0.6	35.8	6.1
Sow (1)	46	11	14	3	32	8
Sow (2)	18.3	5.4	3.2	0.7	15.1	4.7
Growing pig (1)	20	3.85	6	1.3	14	2.5
Growing pig (2)	9.8	2.9	2.7	0.6	7.1	2.3
Layer hen (1)	1.23	0.26	0.36	0.04	0.87	0.22
Layer hen (2)	0.55	0.15	0.05	0.006	0.50	0.14
Broiler (1)	1.09	0.17	0.45	0.075	0.64	0.10
Broiler (2)	0.41	0.11	0.13	0.018	0.28	0.09

1. High production situations
2. Low production situations

To assess the P excreted by monogastrics the same methodology can be applied, but the digestibility of P is less well established, partly because it is influenced by the supply of P in relation to animal requirements, content of other minerals (especially Ca and vitamins in the feed), etc.

In most low production situations, P excreted via urine is low, partly because feeds with low P digestibility are common. In high production situations, P excreted via urine might be considerable (up to 30-40% of total excreted P), especially if most of the feed consists of feedstuffs with a high P digestibility (animal products and mineral P).

Phosphorus excreted via urine by ruminants is usually negligible because a possible surplus of digested P is recirculated to the rumen and from there excreted in faeces instead of in urine.

For most other nutrients, digestibility and retention are not well enough understood to make generalized assessments for the distribution between urine and faeces. It is known, however, that most of the excreted K is in urine.

Species	% of total excreted N	
	Faeces	Urine
Dairy cow 1)	31	69
Dairy cow 2)	50	50
Sow 1)	27	73
Sow 2)	36	64
Growing pig 1)	22	78
Growing pig 2)	41	59
Laying hen 1)	18	82
Laying hen 2)	30	70
Broiler 1)	17	83
Broiler 2)	40	60

1. High production situations
2. Low production situations

AMMONIA EMISSION

The mineral N fraction varies from 0.10 for farmyard manure to 0.94 for urine depending on the type of animal and the type of collection and storage system of the manure. Practically all N in fresh urine is in the form of urea, $CO(NH_2)_2$, which is hydrolysed into ammonium carbonate in the presence of urease. This is an unstable compound and decomposes readily into NH_3 and CO_2. Volatilization of NH_3, however, is governed by the chemical equilibrium between NH_4^+ and NH_3 and the equilibrium between aqueous and gaseous NH_3.

$$CO(NH_2)_2 + H_2O \xrightarrow{urease} (NH_4)_2 CO_3$$

$$(NH_4)_2 CO_3 \xrightarrow{H_2O} 2NH_3 \uparrow + CO_2 \uparrow$$

$$NH_4^+ + OH^- \rightarrow NH_3 + H_2O$$

$$NH_3(aq) \rightarrow NH_3(g)$$

Hydrolysis of urea increases the pH, resulting in higher concentrations of OH in the solution. This forces the equilibrium of Eq. 3 to the right, thus stimulating NH_3production. The rate of volatilization of NH_3 as described in Eq. 4 is linearly related to the vapour pressure of NH_3 in the solution, which is linearly related to the concentration of NH_3 in the solution, and is furthermore temperature-dependent.

7The rate of volatilization also depends on wind speed. The most common method of measuring NH_3 emission in the field is a micro meteorological method described by Denmead (1983).

Adjustment of feeding rations to reduce NH_3 emission from manure

In addition to adjusted manure storage and application techniques, NH_3

emission can be limited by feed ration adjustments. Since most of the N in the urine originates from an imbalance between the amount and the quality of digestible protein and animal requirements, the amount of N in the urine can be reduced by adjusting the feed ration.

In intensive ruminant production systems, balancing the intake of rumen degradable protein and rumen degradable energy can produce major effects. In most European countries this would require a substantial reduction in the quantities of grass fed, as grass with a reasonable productivity per hectare, almost inevitably has a high surplus of rumen degradable protein. However, this would imply a significant shift in feeding practices and even in the whole livestock system.

In intensive monogastric production systems, reductions in N losses will be mainly achieved by adjusting the protein content of the feed to the variable requirements of animals of various age and productivity (phase-feeding) and by balancing amino acid requirements and digestible amino acids offered. For the last-mentioned option, addition of synthetic amino acids is an important and increasingly popular strategy. Many commercial compound feeds for monogastrics contain added synthetic amino acids, mainly lysine and methionine as these are often the first limiting amino acids. High costs may prevent broader application of other limiting amino acids.

Emission from stables and storage

Much research has been conducted on NH_3 volatilization in intensive livestock systems in the Netherlands where the animals are kept indoors for most of the year. Urine and faeces are collected together through the slatted floor in the stables and the manure is stored in liquid form. A considerable part of the total N excreted volatilizes in the stables and during storage.

Some of the data collected in the Netherlands. Depending on the measures taken to reduce NH_3 emission, between 5 and 35% of the total N excreted volatilizes even before it is applied to the land. Measures to reduce the NH_3 emission are based on one or more of the following strategies:

- Minimize the area of contact between manure and air, e.g. by better or more frequent cleaning of the stables and by covering the manure storage;
- Decrease the concentration of NH_3 (and thus the vapour pressure) in the manure by dilution with water;
- Lower the pH in the manure by addition of acid; and
- Stop the formation of NH_3 from uric acid in poultry manure by drying until the dry matter content exceeds 70%.

No data can be found for intensive livestock systems in other climatic zones. It can, however, be envisaged that emissions would be higher in climates with higher average annual temperatures because of the effect of the temperature, and because more aeration of the stables would be necessary.

In locations where there is a concentration of cattle (such as in feedlots, kraals and bomas) faeces and urine are trampled by the animals, this leading to increased NH_3emission. No quantitative data for these situations have been found. Minimizing the contact area between manure and air by cleaning and covering reduces emissions.

Source	Emission per type of animal (as fraction of mineral N in excreta)					
	Dairy cattle	Veal	Pigs	Sows	Layers	Broilers
Stable - normal	0.18	0.19	0.23	0.27	0.11	0.13
Stable with minor adjustments			0.18	0.15		
Stable with major adjustments	0.09		0.12	0.12	0.05	0.04

Adapted from: Van der Hoek, 1994 and Monteny, 1991.

The effect on NH_3 emission of the use of organic materials with a high C:N ratio as bedding material is also worth investigating as it may cause microbial immobilization of N from excreta. Little quantitative information is available, partly due to measurement problems. In one deep litter stable with straw bedding, NH_3 emission was 23% less than the average emission from dairy cattle stables in the Netherlands. Methane emission from this deep litter stable, however, was three to four times higher than the values mentioned in literature for other stables.

Manure storage outside the stables is another source of NH_3 emission. High NH_3 emission will occur from lagoons as the major part of the N is in mineral form, though the level is greatly dependent on factors like depth of the lagoon, storage time and mineralization rate. Schulte (1993) mentions losses of 70-80% of the total N.

In the Netherlands, emission from slurry from dairy cattle stored in separate storage tanks is estimated at 7-14% of the mineral N in the excreta and from pig slurry at 10-19%. Covering the manure storage achieves an emission reduction of 80%.

EMISSION AFTER MANURE APPLICATION TO LAND

The *rate* of NH_3 volatilization is strongly influenced by NH_3 concentration in the manure, temperature and wind speed. These factors, however, do not necessarily influence the *total amount* of NH_3 volatilized during, say, a period of seven days after surface spreading. Hence, the total amount of NH_3 volatilized in seven days in warm, dry and windy conditions may not be very different from that in cool and quiet conditions. The difference is that in the first situation 95% may volatilize during the first day after application and the remaining 5% during day 2-7, while in the second situation volatilization is spread more evenly over the week.

According to various sources, NH_3 volatilization after surface spreading of liquid cow or pig manure amounts to between 30 and 100% of the total mineral N applied. A substantial reduction in NH_3 volatilization after application is possible by using one of the following techniques:

- Dilution of the manure. Experiments with 1:1 and 1:3 water diluted manure have shown substantial reductions in emission.
- Rain. An artificial shower of 20 mm produced by means of a sprinkler installation.
- Immediate tillage after manure application in such a way that the manure is worked in.
- Sod manuring in grassland. The liquid manure is applied in very narrow furrows made in the sward.
- Injection. The manure is injected into the soil at a depth of 10-20 cm.
- "Drag-hose machine". This machine applies the manure very close to the surface in such a way that contact between manure and air is minimal at the time of application.

Application technique[1]	Average	Range
Surface spreading	0.59	0.29-0.91
Diluted manure 1:3	0.24	0.12-0.35
Rain 20 mm	0.15	0.11-0.18
Immediate tillage	0.17	0.14-0.20
Sod manuring	0.06	0.04-0.08
Injection	0.04	0.00-0.13
Drag-hoses	0.35	0.25-0.44

Note:

[1] For an explanation of the application techniques.

Sophisticated machinery has been developed in recent years for the last three options.

Gross estimates of NH_3 emission under the various manure application practices. As the data collected by Monteny (1991) show large variations, the range in measured emission is also given. These variations are attributed to the variable weather conditions during the period of measurement.

EMISSION FROM GRAZED PASTURES

Emission of NH_3 from grazed pastures can be attributed mainly to volatilization from urine patches. Volatilization from dung pats is less significant. Various authors (Denmead et al., 1974; Galbally et al., 1980; Healy et al., 1970; Vallis et al., 1982; Vertregt and Rutgers, 1987) have reported NH_3 volatilization from urine patches ranging from 4 to 29% of the N in the urine. There seems to be a consensus that in temperate climates about 10% of the N in the urine volatilizes and in subtropical and tropical climates the figure is about 25%.

EMISSION FROM FLOODED FIELDS

Terman (1980) reports data on NH_3 volatilization from flooded rice soils ranging from 3-50% when mineral fertilizers such as ammonium sulphate, ammonium nitrate or urea are applied. No data have been found where in flooded conditions, NH_3 volatilization from mineral fertilizers is compared with NH_3 volatilization from animal manure. Incorporation of green manure, however, resulted in a cumulative NH_3 loss of 3-4% whereas loss from incorporated urea was 10%. One can reason that, under flooded conditions, NH_3 loss from animal manure should be lower than loss from mineral fertilizers because (1) the concentration of NH_3 is usually lower after use of animal manure; and (2) decomposition of organic material from manure increases the CO_2 pressure and the increased acidity created reduces NH_3volatilization. This assessment is in agreement with results from Japan where N losses from inorganic fertilizer plots were much higher than that from organic manure plots (51.4 and 22.3 kg per ha respectively), though no distinction was made between volatilization and denitrification.

THRESHOLD VALUE

The main sources of NH_3 in the atmosphere, in countries with intensive livestock systems, are surface spreading of manure and the livestock housing system.

The negative effects of NH_3 volatilization are bad odour and, after deposition, nitrogen enrichment and acidification of the soil and surface water. It is impossible to set general threshold values for NH_3 emission of livestock because of several reasons:

1. Other sources, like industry, cars and households, also contribute to total acid deposition. In most situations NH_3 is less important than SO_2 and/or NO_x. In Europe (including Russia), for instance, NH_3 only contributed to ca. 24% of the total acid emission, compared to more than 50% by SO_2. Only in relatively small areas with high animal densities, like the Netherlands, is NH_3 the major contributor to total acid deposition, though there are indications that NH_3 is also becoming important in other areas.
2. The threshold value is dependent on the type of soil and vegetation, the main problem (vegetation change or N leaching, etc.). In the Netherlands this gives a variation of maximum deposition values of ca. 460% (ranging from 650 to 3000 mol/ha; Heij and Schneider, 1995). Lowest values are mainly related to the increase of nitrophic plants.
3. A threshold value for NH_3 deposition is not easily translated into a threshold value for emission as the dispersion patterns are highly

dependent on climatic factors, such as prevailing wind speed and direction, rain, etc. However, most of the emitted NH_3 is deposited at a short distance from the source.

4. Maximum N deposition is dependent on N uptake via plants. Higher plant uptake reduces the potential acid function of NH_3. Maximum deposition values of NH_3 all assume a zero nett uptake, which is only relevant in non-productive nature areas.

Distance (m)	Vegetation Forest	Other
20	21000	10500
50	7110	3570
100	2340	1170
200	660	330
500	108	54
1000	26.7	13.2

In the Netherlands, the short-term objective is to reduce the total acid deposition (NH_3, SO_2 and NO_x) to 2000 mol per hectare per year or less, a rather arbitrary value.

SURFACE WATER POLLUTION

Direct discharge

In some countries with intensive livestock systems (particularly Eastern Europe), manure is discharged directly into the surface water, often after first having been treated in lagoons. This implies complete loss of nutrients and organic matter in the manure. Direct discharge into surface water causes an-aerobiosis in the water, because the easily decomposable fraction of the manure will start to decompose immediately using all dissolved oxygen and killing the water fauna and flora. In addition, the load of nutrients in the manure will contribute to water eutrophication and heavy metals will cause toxicity. Where grazing animals have free access to surface water, they can urinate and faecate in it or on the slopes leading to it. Direct discharge or runoff into the water will be the result.

Run off

To some degree, runoff of manure into surface water causes the same problems as direct discharge. The most important sources of runoff are locations where animals are concentrated such as feedlots, stables, bomas and kraals. The manure in these places is often drained and the liquid allowed to flow, directly or indirectly via the soil, into surface water. Runoff from manure applied to the surface of agricultural land can also occur.

Factors affecting runoff are:

- Rate of manure application;
- Precipitation, total amount as well as intensity;
- Infiltration capacity of the soil, which is nil or low for frozen, sun-baked and waterlogged soils; and
- Slope of the soil surface.

Threshold values

The values of directly discharged or runoff manure for oxygen demand, nutrient and (some) heavy metal loads all exceed the respective threshold values for effluents into surface water. The EU has set threshold values of effluents into surface water at 25, 15 and 2 mg/l for BOD, N and P, respectively (Council Directive 91/271/EEC, 1991), while typical BOD values for manure and open yard runoff range between 10,000-30,000 and 1,000-2,000 mg/l respectively. These practices should, therefore, never be permitted.

NITRATE LEACHING

Nitrification and denitrification

Nitrogen from manure may be subject to many transformation processes. Fresh animal excreta does not contain N in the form of NO_3. If manure is not aerated (e.g. by composting), NO_3 will not be formed until after application to the soil.

Transformation of NH_4 into NO_3, called nitrification, is a microbiological process that needs oxygen (O_2). In well-aerated soil, nitrification of NH_4 added to manure may be completed in a few days. In less aerated soil it may take weeks. At low temperatures, the process is very slow. NO_3 can also be formed after decomposition of the organic N compounds of manure.

Transformation of organic N into mineral N is a microbial process called N mineralization. Following NO_3 formation, complete or partial denitrification can take place during which N_2 and N_2O are emitted.

$$NH_4^+ + 2O_2 \rightarrow NO_3^- + 2H^+ + H_2O$$

$$5CH_6H_{12}O_6 + 24NO_3^- + 24H^+ \rightarrow 30CO_2 + 42H_2O + 12N_2$$

$$C_6H_{12}O_6 + 6NO_3^- + 6H^+ \rightarrow 6CO_2 + 9H_2O + 3N_2O$$

As in soils usually aerobic and anaerobic patches or conditions may occur concurrently or alternately (rain showers, irrigation runs), both nitrification and denitrification processes may take place both at the same time and alternately. This makes it extremely difficult to predict the fate of applied nitrogen that is not volatilized or added to the soil N, or recovered by the crop: leached as NO_3 or emitted as N_2/N_2O. An example of an estimate of this 'N.

Input:		mineral fertilizers 350 kg N
	mineral N from slurry	78 kg N
	N mineralized from slurry (30% from organic N)	47 kg N
Total		475 kg N
Output:		N in harvested grass 400 kg N
Balance:		75 kg N

If it is assumed that no immobilization of the remaining N occurs, the balance of 75 kg N must be either leached as NO_3 or denitrified as N_2 or N_2O. This balance is sometimes referred to as "N not accounted for".

Factors affecting fate of "N not accounted for": leaching or denitrification

The level of N fertilization the major factor determining the amount of N available for NO_3 leaching or denitrification. It may be assumed that with moderate N doses (0-150 kg/ha for arable crops or 0-300 kg/ha for grassland in temperate zones) no NO_3 is present in the soil at the end of the growing season. Only higher N doses would result in residual NO_3, for grassland in a cut-and-carry system on clay soil in the Netherlands. High N doses may be applied in situations where animal manure is seen as a waste or where the level of mineral N fertilization is determined without taking into consideration the N effect from animal manure if both nutrient sources are combined.

Under grazed pastures, NO_3 leaching can be considerable even at moderate rates of N fertilization because of leaching from urine patches, which is up to 5 times higher than under grassland without grazing. Van der Meer and Meeuwissen (1989) present the results of a study on the effect of grazing on nitrogen emissions. It is clear that on a well-drained, sandy soil in the Netherlands, NO_3 leaching is much higher under grazed pastures than under a cut-and-carry system. Similar data for other soil or types or climates have not been found.

	N fertilization (kg/ha/year)					
	0	**100**	**200**	**300**	**400**	**500**
N load in urine patches (kg/ha)	225	323	454	595	717	858
NO_3 leaching without urine	15	18	22	28	57	103
Net effect of urine on leaching	2	4	22	58	89	116
Total leaching	17	22	44	86	146	219

Not only the level of N fertilization, but also the timing of application affects NO_3 leaching. Nitrate can accumulate in the soil after removal of the crop through N mineralization from soil organic matter and easily decomposable organic N from crop residues or manure (N_e). The "N not accounted for" must be considered a loss. Many factors influence the fate of the residual N between the end of the growing season and the subsequent crop. The following crop can be sown immediately after or even before harvest of the main crop, thus acting as an "interceptor" of NO_3 that otherwise would have leached or been emitted as N_2/N_2O after denitrification. Without this so-called cover crop, the partitioning between leaching and denitrification is influenced by the rate of NO_3 formation, the balance of precipitation and evaporation, the amount of organic matter in the soil and its drainage conditions.

Leaching from unsealed kraals, feedlots and solid manure storage heaps can be considerable, as the conditions are similar to very high application rates. In general, leaching from lagoons should be less than from solid manure storage heaps as its anaerobic conditions will hardly allow for nitrification. No appropriate quantitative data have been found on this subject, but the level of leaching is highly dependent on the precipitation surplus, in case of uncovered kraals, feedlots and solid manure storage heaps, and on soil conditions (permeability).

Threshold values

A maximum concentration of 50 mg NO_3 per litre is allowed for drinking water by the World Health Organisation (WHO) of the United Nations. At higher NO_3concentrations, public health is at stake through the formation of toxic nitrite from NO_3 in the human body. Babies are particularly vulnerable to nitrite if drinks are made of water exceeding the threshold value. To reach this objective, less than 34 kg N/ha may leach annually, given a precipitation surplus of 300 mm/y (50 mg NO_3 = 11.3 * 10^{-6} kg N; 300 mm precipitation surplus = 3 * 10^6 l/ha; 11.3 * 3 = 34 kg N). A translation into a maximum N surplus per ha, however, is impossible as quantities N emitted via denitrification, ammonia emission, etc. vary widely.

PHOSPHORUS LEACHING

Leaching of P from agricultural land into surface water is an important contribution to its eutrophication. This may lead to excessive growth of algae and eventually to anaerobic water. The water becomes smelly and water life is seriously disturbed.

Due to the complex sorption and fixation behaviour of soils with regard to P, no direct relation has been found between P fertilization and P leaching. For sandy soils in the Netherlands, however, a relation can be given between P in the soil solution, P saturation level of the soil and P leaching. The P saturation level indicates to what extent the soil's P retention capacity has been utilized. When P retention capacity of the soil body above the highest ground water level exceeds 25%, it becomes P saturated and P-leaching to ground water and surface water is almost inevitable even if fertilization is reduced. This is one of the reasons why governments place much emphasis on the containment of P_2O_5 over-fertilization.

Phosphorus balance

For agricultural purposes, some over-fertilization is necessary, i.e. more P should be added to the soil than removed in crop products to maintain an adequate P condition of the soil. The required overdose of P depends on factors such as crop, soil type and ground water level, but is in any case 25 kg P_2O_5/ha/year or more in the Netherlands. Legislation in the Netherlands on the use of animal manure in agriculture is based on the addition to the soil of P_2O_5. The period 1990-2000 the maximum amounts of P_2O_5 in animal manure that have been and probably will be allowed to be added to the soil for different crops. As it was common practice for many intensive livestock holders to use maize plots as "dumping sites" for their manure surpluses, the law allowed for relatively high amounts of P_2O_5 on maize during the first years, to give those farmers the opportunity to adjust their manure management to the new legal situation. The objective is that by the year 2000 the estimated P_2O_5 addition to the amount removed in crops will be restricted to 90 kg/ha for grassland and 65 kg/ha for maize and other field crops averaged over the country. As yields vary among regions and farms, so will crop removal of P_2O_5. The present idea is, therefore, to identify farm-specific maximum doses of P_2O_5application. A compulsory P balance registration system at farm level will be the tool to determine this dose. Whether legislation and law enforcement can manage this system is questionable.

	1990	1991-2	1993	1994	1995	1996	2000
Grass-land	250	200	200	200	150	135	90
Maize	350	250	200	150	110	90	65
Field crops	125	125	125	125	110	90	65

Note:

Legislation is given in kg P_2O_5, which is 2.29 x P

Currently, policy makers and specialists are discussing specification of an acceptable overdose on top of the amount of P_2O_5 removed by the crop.

A large gap, however, exists between the environmentally desirable overdose and the agriculturally optimal overdose (>25 kg/ha). Discussion on this topic has been going on for some years and is expected to continue. The basic principle of balancing P addition to the soil and P removal in crops will be maintained.

Phosphorus balance in relation to other environmental problems of manure application

It is sometimes stated that environmental problems like N leaching and heavy metal accumulation will be negligible if P equilibrium is maintained at field level. This statement is obvious for N leaching because the N:P ratio in manure is in all but very exceptional cases lower than the N:P ratio in crops. Thus, N removal in crops will be generally higher than N-application via animal manure

The exception is Cu, as the Cu content of pig manure remains relatively high, even though Cu additions to pig feed in the European Union have been reduced significantly in the recent past: from 200 mg per kg 125 mg in 1978 to 175 and 35 mg at present in starter and fattening feed respectively.

Continuous application of other types of animal manure will cause even less problems as heavy metal concentrations in the feed of these animals are lower, with the possible exception of Cd in broiler feed (Heidemij, undated). Note, on livestock farms, when buying part of the feed as concentrate, heavy metal accumulation will always be possible as both retention in animal products and leaching are very low, resulting in a gradual build up of heavy metal concentrations in home-grown feed.

Adjustment of feeding rations to reduce P excretion

Because P in manure plays such a pivotal role in the Dutch manure legislation, much research is being conducted at present on ways to reduce P excretion of livestock. Until recently, fairly large safety margins for P requirements of livestock were used, as more accurate assessment of these requirements is complicated and no problems were anticipated in case of overdose. Now more research has been started, particularly on monogastrics, to establish more accurate information on digestible P requirements and P digestibility for various feeds.

It is commonly assumed that P digestibility of feedstuffs of plant origin is about 30-35%, but large variations seems to exist, ranging from 20% or lower for maize and rice bran, to ca. 38% for barley and soybean meal and ca. 46% for wheat and peas. Digestibility is highest in feedstuffs of animal origin, ranging from 70 to 90%.

Part of the variation in P digestibility of feeds of plant origin can be explained by the content of both phytate P and phytase: phytate P is hardly digested, but if phytase activity is high P digestibility will increase. Phytase of plant origin is present in germinating seeds, but also in grains in rest, particularly rye and wheat. Major efforts have been made to examine the effects of the addition of microbial phytase to pig and poultry feeds to increase P digestibility, and consequently, to reduce the amount of P excretion. It is estimated that P digestibility can be increased up to 60% via the use of phytase. Important aspects of research were the thermo-resistance of the enzyme, as temperatures above 80 °C occur during pelleting, and its activity under low pH conditions resulting in the identification of micro-organisms producing suitable enzymes for use in the compound feed industry.

An alternative possibility to reduce P excretion is to alter feed rations to include ingredients with higher P digestibility i.e., mainly feedstuffs of animal origin or leguminous feeds. However, the use of more leguminous feeds is at this moment seriously hampered by the occurrence of high levels of anti-nutritional factors in most of these feeds.

Threshold values

The policy in the Netherlands aims at a total P concentration in surface water below 0.15 mg per litre. Research has indicated that at this concentration, no excessive growth of algae will occur. The water will remain sufficiently aerobic to be suitable for multiple use: fish, water recreation, irrigation, etc.

To reach the environmental objective of 0.15 mg P total/l or less in the ground water, the P_2O_5 overdose in a possible steady state situation in the Netherlands cannot exceed 1 kg P_2O_5/ha/yr (The annual precipitation surplus in the Netherlands is 300 mm, which is $3*10^6$ l per ha; this volume of water with P concentration of 0.15 mg/l contains $3*10^6 * 0.15*10^{-6} = 0.45$ kg P; this equals $0.45 * 2.29 = 1.03$ kg P_2O_5). At an annual precipitation surplus of, say, 600 mm, the environmentally maximum allowable overdose would be 2 kg P_2O_5/ha.

For ground water, a threshold value of 0.10 mg ortho-P/l is suggested for sandy soils at the depth of the mean highest ground water level. At this concentration, flow of ground water to surface water is assumed not to cause the concentration of P in the surface water to exceed the indicated threshold value.

AGRICULTURAL VALUE

The fertilizing value of animal manure not only depends on its composition, but also on the crop to which it is applied, the climatic conditions, the type of mineral fertilizer substituted, the method of application and the time of application in relation to crop growth.

The efficiency of a nutrient in animal manure can be expressed in the working coefficient. The working coefficient indicates how much nutrient from a reference mineral fertilizer is used to produce the same yield as an application of animal manure which contains 100 kg of the nutrient. For both animal manure and mineral fertilizer, nutrient-yield response curves can be plotted. Usually the crop responds better to nutrients in a mineral fertilizer than to nutrients in animal manure.

Fertilizing value of nitrogen in manure

As mentioned earlier, N in manure and in soil can be present in various forms. Furthermore, plants and microorganisms can use N in different ways. Because of this complexity, crop utilization of N is rather unpredictable. It would be best to establish, through experiments, crop response curves under all circumstances relevant to the identified livestock systems and determine the working coefficients as shown above. Only through experimentation can working coefficients. Such data, based on experiments, are often not available, so an alternative approach is suggested.

First, the form in which N is present in the manure has a decisive influence on the availability for plants. Higher plant availability of N in the manure implies that it can substitute for more mineral fertilizer, i.e. the working coefficient will be higher. Sluijsmans and Kolenbrander (1977) suggested differentiation between:

- N_m: N present in mineral form;
- N_e: N mineralized within one year after application in temperate zones or within three months in lowland tropics;
- N_r: N mineralized later.

The fractionation depends, among other things, on the type of animal, its diet and the method and duration of storage of the manure. The working coefficients can now be estimated with some assumptions.

These include:

	N_m	N_e	N_r
Cattle farmyard manure	0.10	0.45	0.45
Cattle liquid manure	0.50	0.15	0.35
Pigs liquid manure	0.50	0.22	0.28
Poultry liquid manure	0.70	0.20	0.10
Veal liquid manure	0.80	0.09	0.11
Urine	0.94	0.03	0.03

1. Ammonia volatilization during and after application: One of the processes determining the working coefficient of N_m is volatilization of NH_3. The working coefficient for N_m can be considered as (1-x), where x is the fraction of volatilized N_m. When x=1, all N_m volatilizes as NH_3, which may happen when liquid manure is surface applied without being worked in.

   Flooded rice is a special case. Here, the manure mineral N dose will usually be low compared with the urea mineral N dose. In both cases the mineral N is diluted many times by the irrigation water. Decomposition of manure, however, will increase the CO_2 concentration in the water and decrease the pH. Therefore, x will be very low. The NH_3 volatilized from urea will be more than that from the reference fertilizer.
2. NH_3 volatilization before application: To estimate the mineral N (practically all in the form of NH_3) *at the time of application* N_m, volatilization *before*application, in the stable and during storage should be taken into account.
3. NO_3 leaching: The wheat crop only utilized half of the N mineralized annually. The remainder is assumed to have leached out or denitrified. Timing of the manure application is important in

this respect. A longer period between manure application and crop planting date leads to higher emissions. Autumn application, winter fallow and spring planting, would result in a much lower N working coefficient.

4. Temperature: Higher temperatures result in more rapid N mineralization. As rule of thumb Janssen (1993) says that the decomposition rate doubles for every 9 °C rise in average annual temperature. N_r to the temperate zone at an average annual temperature of 9 °C. For a climate with an average annual temperature of 18 °C, N_r refers to organic N six months after application.

For tropical lowland at an average annual temperature of 27 °C, N_r refers to residual organic N three months after application of the manure.

Applying the indicated calculation method and the data is possible to calculate for a particular livestock system how much fertilizer N can be substituted by N from animal manure, provided manure production is known.

In a very high rainfall situation the recovery of N (kg N in crop/kg N applied) from manure may be higher than from inorganic fertilizer, due to the slow release character of manure resulting in N mineralization which is better synchronized with plant uptake than inorganic fertilizers. This would result in a N working coefficient of manure, in terms of fertilizer equivalents, higher than 1.0 minus 'before application losses'.

Under flooded conditions N recovery is low for both inorganic and organic fertilizer: estimated at 35 and 25% respectively from a long-term experiment in Japan. However, these results were obscured by confounding effects, such as highly reduced biological N fixation and decreasing soil N reserves in the case of inorganic fertilizer. If these effects were included, N utilization from organic manure would be much higher than from inorganic fertilizer. Hence, N working coefficients would again be higher than 1 minus 'before application losses'.

Fertilizing value of phosphorus in manure

For practical recommendations to Dutch farmers, Noij and Westhoek (1992) differentiate between a single dose and repeated annual doses of P. They assume a working coefficient of P from any type of slurry on grassland of 0.8 after a single dose and of 1.0 after repeated annual doses. For maize, they assume a working coefficient for cattle, pig or poultry slurry of 0.6, 1.0 and 0.7, respectively for a single dose, and again 1.0 if repeated annual doses are applied. For maize, a fraction of the P is released only in later years. Working coefficients for P from farmyard manure higher than 1.0 have been reported from experiments on potatoes over a period of six years in India. Working coefficients for P from farmyard manure of 0.6, 0.4 and 0.5 were from experiments on hyacinth bean (*Dolichos lablab*L.) for three consecutive seasons Noor *et al.* (1992), An indicative value for the working coefficient of P for any type of animal manure is 1.0 with repeated annual applications. For single applications of farmyard manure, the P working coefficient can be assumed to be 0.5.

Temperature influences mineralization of the organic P in manure. It can therefore be assumed that like N, P is mineralized more rapidly in tropical than in temperate conditions. Because the ratio organic P/inorganic P in manure is much lower than organic N/inorganic N, temperature influences the working coefficient of P for single applications much less than that of N.

Fertilizing value of potassium in manure

The sources quoted suggest a working coefficient of 1.0 for K in animal manure. Noij and Westhoek (1992), however, make an exception for animal manure applied in the Netherlands on sandy soils before February, in which case leaching of K may occur, which will reduce the working coefficient to 0.8 on average.

Contribution of manure to soil organic matter

In clay soils, soil organic matter is important for soil porosity and aeration. On loamy and sandy soils, soil organic matter increases water holding capacity, cation exchange capacity (CEC) and improves soil structure. A higher soil organic matter content, therefore, makes soils less susceptible to erosion. Any addition of organic material, including manure contributes to soil organic matter content. This humification coefficient indicates which fraction of the added organic material remains after a certain period of decomposition as "stable organic matter" and, thus, contributes to soil organic matter. Manure has a relatively high humification coefficient, so the contribution of animal manure to the build up of soil organic matter is, therefore, high. However, for the production of one ton dry matter as manure, more dry matter as fodder is needed. Livestock systems therefore, may not contribute more to soil organic matter than do agricultural systems without livestock. On the other hand, livestock systems often involve cropping patterns with crops that contribute more to soil organic matter content than arable crops. Only in situations where potential feed biomass is wasted can livestock play a positive role in this respect.

THE VALUE OF MANURE

Manure is often an under-valued resource. When well-managed and properly applied, it reaps many benefits. The benefits can be seen in the photo to the right, but are not captured in much of the field because of errors in application.

- Manure as a nutrient source can be a substitute for purchased fertilizers

- Manure increases crop productivity and can be used in gardens and flowerbeds, not just agriculture.
- The organic matter in manure improves water infiltration into the soil and reduces runoff and erosion. For this reason, when manure replaces synthetic fertilizers at agronomic phosphorus rates, there is often reduced risk of phosphorus runoff.
- Some manure types have a liming effect and reduce soil acidity.
- The P-Index rating is often reduced.

Manure tends to reduce sediment and water loss from a field, but the concentration of phosphorus in the runoff is typically high when soil test P is high. Manure application may, therefore, result in either increased or decreased phosphorus runoff. Phosphorus runoff loss may be reduced with manure application due to reduced erosion and runoff in cases of agronomically moderate application rates and moderate soil test P. The combination of land-applying higher solids-manures and good conservation best-management-practices can significantly reduce erosion and runoff and maintain phosphorus losses at an acceptable level. However, if soil test P is excessively high, the reductions in erosion and runoff due to manure application are often not sufficient to prevent unacceptable losses of P.

*Cattle manure as fertilizer: One ton contains approximately 12 pounds of nitrogen. About 25% of this is immediately available to the crop. So, if 35 tons of cattle manure were spread over an acre of land, the available rate would be approximately 105 pounds of N for immediate crop use. If that amount of nitrogen had to be purchased, it could cost more than $68 per acre at 2008 prices ($0.65/lb), and this does not account for the other nutrients found in manure or the amount of nitrogen available to the crop in years to come. Phosphorus is also found in cattle manure. If you had to apply phosphorus as well as nitrogen, it could cost an additional $127.40 per acre if applied at the rate equivalent to the available amount (2.6 lb/ton of manure) in the manure using projected 2009 prices ($1.40/lb). In addition to nitrogen and phosphorus, manure contains zinc and sulfur, also important nutrients for crops.

MANURE IS AN EXCELLENT FERTILIZER

When developing a manure by-product market is important to understand that manure is a necessary by-product of the livestock industry and it is the technology involved in the treatment system that determines whether manure is a valuable resource or a costly liability.

The simple fact is, untreated manure is simply animal feces while properly treated/processed manure is a value added marketable organic residual. Technological factors involved in manure treatment systems have a significant influence on by-product quality and it is the quality that dictates the value.

In addition to agriculture, the potential markets for high quality, composted manure products include horticulture i.e. gardening, landscaping, nurseries, topsoil production - silviculture i.e. Christmas trees, ornamentals - reclamation i.e. landfill covers, mine reclamation and other environmental uses i.e. biofilters, erosion control and wetlands restoration to name a few.

MANURE AS A FERTILIZER

Manure is an excellent fertilizer containing nitrogen, phosphorus, potassium and other nutrients. It also adds organic matter to the soil which may improve soil structure, aeration, soil moisture-holding capacity, and water infiltration.

To determine how much manure is needed for a specific application, the nutrient content and the rate nitrogen becomes available for plant uptake needs to be estimated. Nutrient content of manure varies depending on source, moisture content, storage, and handling methods.

Nitrogen content in manure varies with the type of animal and feed ration, amount of litter, bedding or soil included, and amount of urine concentrated with the manure. Moisture content is also a major consideration. For example: The moisture content of fresh manure is around 70% to 85%. The moisture content of air-dried manure is around 9% to 15%. As manure dries, the nutrients not only concentrate on a weight basis, but also on a volume basis due to structural changes (settling) of the manure. Volatilization of urine nitrogen can result in considerable loss of nitrogen, up to 50% or more of the total nitrogen.

Generally, dry manure contains 1.5 to 2.2 cubic meters per ton. Dry poultry and steer manure contain around 1.9 cubic meters per ton.

MANURE HANDLING

Handling can affect the fertilizer value of manure, particularly its nitrogen content. Nitrogen is present in manure in a variety of forms, most of which gradually converts to ammonium and nitrate nitrogen.

The ammonium form can be lost to the air and the nitrates leached by rainfall. Ammonium losses can be minimized by not stockpiling manure while it is moist, minimizing its handling, and working it under immediately after spreading. Ammonia can be lost to the air each time manure is moved or hauled. Much of the loss is from hydrolysis of the NH_2 groups (enzymatic) and then volatilization of N_20 and NH_3. This loss can be very high when spreading manure, especially during warm, dry weather. Here, at least 50% of the ammonium nitrogen can be lost within 12 hours. Studies have also shown that, by one week after spreading, almost 100% of the ammonium nitrogen can be lost. This loss can represent up to 50% of the total nitrogen available in stockpiled manure.Therefore, the importance of simultaneously spreading and working in manure is obvious.

Nutrient Availability and Manure Application

Manure is a source of many nutrients including: nitrogen, phosphorus, potassium and many others. However, nitrogen is often the main nutrient of concern for most crops. Potassium deficiency is usually quite localized within a field and would not be corrected with common rates of manure. However, some improvement might be expected with high rates above 10 tons per acre. The high rates needed to correct a potassium (K) deficiency would supply an excess amount of nitrogen for many crops, and this should be avoided.

	Nitrogen	Phosphorus	Potassium	Calcium	Magnesium	Organic matter	Moisture content
	(N)	(P_2O_5)	(K_2O)	(Ca)	(Mg)		
FRESH MANURE	%	%	%	%	%	%	%
Cattle	0.5	0.3	0.5	0.3	0.1	16.7	81.3
Sheep	0.9	0.5	0.8	0.2	0.3	30.7	64.8
Poultry	0.9	0.5	0.8	0.4	02	30.7	64.8
Horse	0.5	0.3	0.6	0.3	0.12	7.0	68.8
Swine	0.6	0.5	0.4	0.2	0.03	15.5	77.6
TREATED DRIED MANURE	%	%	%	%	%	%	%
Cattle	2.0	1.5	2.2	2.9	0.7	69.9	7.9
Sheep	1.9	1.4	2.9	3.3	0.8	53.9	11.4
Poultry	4.5	2.7	1.4	2.9	0.6	58.6	9.2

Rates of Manure for Nitrogen Needs

The nitrogen compounds in manure are eventually converted to the available nitrate form. Nitrate is soluble and is moved into the root zone with water. It is the same form ultimately available to plants from commercial nitrogen fertilizers.

However, the release of available nitrogen from the complete organic compounds during manure decomposition is very gradual. This slow release of nitrogen is manure's most important asset. It extends nitrogen availability and reduces leaching — of particular importance in sandy soils.

The idea is to first apply enough manure to meet the first year's need of available nitrogen. Decreasing amounts are then applied in following years because of the carry-over organic nitrogen that will be released from previous applications.If the same rate of manure is applied each year, it is possible for a field originally low in nitrogen to accumulate unnecessarily high levels in successive years.

The nitrogen in poultry manure is in released fastest, about 90% is released in the first year.

Fresh manure which contains both the urine and solid portions and has a large amount of urea or uric acid provides a somewhat slower release rate, with approximately 75% of the total nitrogen released the first year.

An even more gradual nitrogen release can be expected from dry feedlot steer manure, with only 35% of the total nitrogen released the first year.

OTHER BENEFITS OF MANURE

The use of manure helps to maintain the organic matter content of the soil which can improve soil structure and water infiltration. However, manure is quickly decomposed under warm, moist soil conditions. With the manure rates used for most crops, organic matter content in soil is only temporarily increased.

POSSIBLE DISADVANTAGES

Weed seeds are common in some manure. They may enter the animal with its feed and then pass through the digestive tract, still viable, or they may have come with the litter, or they may have simply blown into the feed yard.

Poultry droppings typically have fewer weed seeds surviving the digestive processes. However, other animal manure may contain numerous viable weed seeds if the original feeds were contaminated. Composting and stockpiling manure can reduce the number of viable weed seeds.

Manure commonly contain 4 to 5% soluble salts (dry weight basis) and may run as high as 10%. To illustrate, an application of 5 tons of manure containing 5% salt would add 500 lbs. of salt.

Normally, irrigation and rain water will sufficiently leach well-drained soils to prevent damaging salt accumulations. However, one should be cautious with poorly drained soils, soils with existing salinity problems, or unusually high application rates, especially when concentrated near young plants.

Zinc deficiency can be induced or increased with repeated high rates of manure, especially on sandy soils.

Moderate or infrequent applications do not normally present a zinc problem. However, growers should be aware of the potential problem, especially with soils and varieties or crops of known susceptibility to zinc deficiency.

Summary

The principal value of manure is its extended availability of nitrogen — of particular value in the more readily leached sandy soils. Manure is also helpful in improving soil fertility in cut areas from land leveling.

Nutrient content and rate of availability varies widely, depending mostly on manure source, handling methods, and water content. Fresh manure which includes both liquid and solid fractions with the least handling and then work in immediately after spreading will retain the most nitrogen. A laboratory analysis of the manure for nitrogen content is useful. An accurate sample of the manure requires a composite of many samples throughout the pile or lagoon.

Generally, poultry manure is highest in nitrogen content, followed by hog, steer, sheep, dairy, and horse manure. Feedlot, steer manure requires fairly high rates to meet first-year nitrogen requirements because of its lower nitrogen percent and gradual nitrogen release characteristics.

However, this feature provides for more continued nitrogen availability in succeeding years, allowing for progressively lower annual application rates to meet plant requirements.

Faster nitrogen-release sources, such as poultry manure, require more constant and lower annual rates to maintain nitrogen availability.

The possible advantages of organic matter content and disadvantages of weed seed and salt content should be considered in using manure.

MANAGING MANURE FERTILIZERS IN ORGANIC SYSTEMS

The basics for manure management in organic systems. Topics covered include National Organic Program regulations, the risk of contaminants in manures, guidelines on how to manage nutrients in manure, and testing manure or compost. Some of the challenges of nutrient supply and test interpretation associated with the repeated use of manures are discussed along with tips and tools you might use to determine manure application rates.

INTRODUCTION, RULES, AND CONCERNS

Livestock manure is a key fertilizer in organic and sustainable soil management. Manure provides plant nutrients and can be an excellent soil conditioner. Properly managed manure applications recycle nutrients to crops, improve soil quality, and protect water quality. It is most effectively used in combination with crop rotation, cover cropping, green manuring, liming, and the addition of other natural or biologically-friendly fertilizers and amendments.

Use of manure imported from conventional farming operations is allowed by National Organic Program (NOP) standards. There are, however, application restrictions. Manure may only be used in conjunction with other soil-building practices and be stored in a way that prevents contamination of surface or ground water. Many certifiers specify that manure application must not exceed "agronomic application rates", which means the amount applied must be less than or equal to the requirements of the crop. Manure cannot be applied when the ground is frozen, snow-covered, or saturated.

The NOP regulation specifies that "raw" fresh, aerated, anaerobic, or "sheet composted" manures may only be applied on perennials or crops not for human consumption, or such uncomposted manures must be incorporated at least four months (120 days) before harvest of a crop for human consumption, if the crop contacts the soil or soil particles (especially important for nitrate accumulators, such as spinach). If the crop for human consumption does not contact the soil or soil particles (e.g. sweet corn), raw manure can be incorporated up to 90 days prior to harvest. Biosolids, sewage sludge, and other human wastes are prohibited. Septic wastes are prohibited, as well as anything containing human waste.

Composted plant and animal manures are those that are produced by a process that: (i) established an initial C:N ratio of between 25:1 and 40:1; and (ii) maintained a temperature of 131°F to 170°F for 3 days using an in-vessel or static aerated pile system; or (iii) a temperature of between 131°F and 170°F for 15 days using a windrow composting system, during which period, the materials must be turned a minimum of five times. Alternatively, acceptable composts must meet the November 9, 2006 NOSB Recommendation for Guidance Use of Compost, Vermicompost, Processed Manure and Compost Tea that identifies materials and practices that would be acceptable.

Heat-treated, processed manure may be used as a supplement to a soil-building program, without a specific interval between application and harvest. Producers are expected to comply with all applicable requirements of the NOP regulation with respect to soil quality, including ensuring the soil is enhanced and maintained through proper stewardship.

According to the NOP's July 17, 2007 ruling, "processed manure products must be treated so that all portions of the product, without causing combustion, reach a minimum temperature of either 150°F (66°C) for at least one hour or 165°F (74°C), and are dried to a maximum moisture level of 12%; or an equivalent heating and drying process could be used." To achieve equivalency status, processed manure products can not contain more than $1x10^3$ (1,000) MPN (Most Probable Number) fecal coliform per gram of processed material sampled and not contain more than 3 MPN Salmonella per 4 gram sample of processed manure.

As always, organic vegetable growers should get label information and check with their certifiers before using purchased compost or processed manure products.

Some manures are contaminated with hormones, antibiotics, pesticides, disease organisms, heavy metals, and other undesirable substances. Many of the organic compounds, pathogens, protozoa, or viruses can be eliminated through high-temperature aerobic composting. Caution is advised, however, as some disease causing agents, e.g. *Salmonella* and *E. coli* bacteria, may survive the composting process. Manure and compost testing is available through commercial labs and is recomended in situations where there is any doubt about the purity of manures. Manure testing is required by the European Union and Canadian standards. The possibility of transmitting human diseases discourages the use of fresh manures and even some composts as pre-plant or sidedress fertilizers on vegetable crops. Apply animal manures at least 90 or 120 days, as applicable, prior to harvest of any crop that could be eaten without cooking.

Best management practices recommended for manure are as follows:

1. Avoid manuring after planting a crop to be harvested.
2. Incorporation before planting is recommended.
3. Do not use dog or cat (fresh or composted) because these species share many parasites with humans.
4. Wash all produce from manured fields thoroughly before use.

Cautions or concerns include the following:

1. Manures imported from conventional farms can contain residues from hormones or pesticides.
2. In rare cases, carryover of persistent herbicides can occur. Most herbicides break down rapidly after application or during normal composting. However, some of those in the pyridine carboxylic acid group such as clopyralid, which is commonly used on grass lawns, break down slowly, even during composting, and are not degraded when ingested by animals because they pass into the urine quickly. Application of manures or composts derived from grass treated with clopyralid is restricted during the "growing season of application" for all farms, not just those that are organic.
3. Heavy metals (e.g., As, Cu, and Zn) are fed to livestock and then added to soils in the form of manures. Unlike sludge, metal content does not influence manure application rates to soils but should be considered as metals persist in the soil and will accumulate with repeat application. Concerns over heavy metals, other chemical contaminants, and salinity are most often raised in association with poultry litter. Under federal organic standards, certifiers may require testing of manure or compost if there is reason to suspect high levels of contamination.
4. Weed seeds and plant diseases can be effectively controlled by high temperature aerobic composting of manures.

MANURE HANDLING: RAW STACKED OR COMPOSTED

The NOP regulation also requires that manure and other fertility inputs must be managed so that they do not contribute to contamination of crops, soil, or water by excess nutrients, pathogens, heavy metals, or residues of prohibited substances. Whether animals are raised on farm or manures are imported, organic farmers are likely to need to store manure on farm prior to application. Proper manure storage conserves nutrients and protects surface and groundwater. Storing manure can be as elaborate as keeping it under cover in a building, or as simple as covering the manure pile with a tarp. The important point is keeping the pile covered and away from drainage areas and standing water. The storage location should also be convenient to your animals and crop production.

When you are looking for organic forms of nutrients for crop production, manure and manure composts are two of the logical choices. Composting is more than just piling the material and letting it sit. Composting is the active management of manure and bedding to aid the decomposition of organic materials by microorganisms under controlled conditions. Weed and disease problems associated with raw manures can be alleviated with proper composting. Use of composted manures can also reduce P transport to waterbodies.

Organic producers making their own compost must keep records of their composting operation to demonstrate that the compost was produced according to the definition cited above. If the compost is purchased, the grower should ask for documentation from the supplier showing that the compost meets NOP requirements. Keep this documentation, along with purchase receipts, with your other records. If the compost is 100% plant-based, without any animal excrement or by-products, there is no requirement for heating or turning.

Table. Comparison of composted and raw manures.

Compost	Manure
slow release form of nutrients	usually higher nutrient content
easier to spread	sometimes difficult to spread
lower potential to degrade water quality	higher potential to degrade water quality
less likely to contain weed seeds	more likely to contain weed seeds
reduced pathogen levels (e.g. salmonella, E. coli)	potential for higher pathogen levels
higher investment of time or money	lower investment of time or money
more expensive to purchase	less expensive to purchase
fewer odors (although poor composting conditions create foul odors)	odors sometimes a problem
improves soil tilth	improves soil tilth

MANAGING NUTRIENTS IN MANURE

Manure nutrient contents are highly variable and growers must be able to understand and reduce this variability to make the best agronomic and environmental use of these resources. Manure must be carefully managed to prevent over- or under-application and to account for the cumulative environmental effects of application as well as storage. Balancing crop nutritional needs with manures is an ongoing challenge. Finding out about manure composition is critical to its efficient use. Applying too little can lead to inadequate crop growth because of lack of nutrients. Over-application can reduce crop quality and increase the risk of plant diseases. Over-application will also increases the risk of contaminating surface or groundwater.

There are three main sources of variability and uncertainty when using manure:

1. Nutrient and moisture content of the manure.
2. Material heterogeneity and application variability.
3. Availability of nutrients to crops.

Manure composition varies with the species of animal, feed, bedding, and manure storage practices. These values may not accurately represent your situation. Nutrient values can vary by a factor of two or more from the values. This is why it is important to test materials applied instead of guessing.

Table. Typical nutrient content of manure (from Koelsch and Shapiro, 2006). Because of variability between farms, individual manure analysis is preferable to the estimates below.

	% Dry Matter	Ammonium–N	Organic–N	P_2O_5	K_2O
Slurry Manure	(lb. of nutrient per 1,000 gallons of manure)				
Dairy	8	12	13	25	40
Beef	29	5	9	9	13
Swine (finisher, wet-dry feeder)	9	42	17	40	24
Swine (slurry storage, dry feeder)	6	28	11	34	24
Swine (flush building)	2	12	5	13	17
Layer	11	37	20	51	33
Dairy (lagoon sludge)*	10	4	17	20	16
Swine (lagoon sludge)	10	6	16	48	7
Solid Manure	(lb. of nutrient per ton of manure)				
Beef (dirt lot)	67	2	22	23	30
Beef (paved lot)*	29	5	9	9	13
Swine (hood barns)	57	4	13	20	
Dairy (scraped earthen lots)	46	3	14	11	16
Broiler (litter from house)	70	15	60	27	33
Layer	40	18	19	55	31
Turkey (grower house litter)	70			15	30
Liquid Effluent from lagoon or holding pond	(lbs. of nutrient per acre-inch)				
Beef (runoff holding pond)	0.25	71	8	47	92
Swine (lagoon)	0.40	91	45	104	189
Dairy (lagoon)	2	317	362	674	1082

Value based upon ASAE, 2005, D384.2; Manure Production and Characteristics with exception of those marked with an "*".

MANURE SAMPLING AND TESTING

Commercial laboratories can measure the nutrients in manure and save you from guessing based on table values. Testing laboratories typically charge from $30 to $60. It is important to use a laboratory that routinely tests animal manure, as they will know the correct type of analysis to use.

A nutrient analysis is only as good as the sample you take. The best time to sample by far is right before you apply the material because N loss in storage is accounted for. Also, if you use manure repeatedly from the same source, you can develop a running average analysis of that manure (over a 3+ year period). A running average is more likely to be accurate than a single sample taken from a storage pile or lagoon. Samples must be fresh and representative of the manure. Follow these steps:

1. Ask the laboratory what type of containers they prefer and make sure the laboratory knows when your sample is coming. Laboratories should receive samples within 48 hours of collection. Plan to collect and send your sample early in the week so the sample does not arrive at the lab on a Friday or a weekend.
2. If you have a bucket loader and a large amount of manure, use the loader to mix the manure before sampling.
3. Take 10–20 small samples from different parts and depths of the manure pile to form a composite sample. The composite sample should be about 5 gallons. The more heterogeneous your pile, the more samples you should take.
4. With a shovel or your hands thoroughly mix the composite sample. You may need to use your hands to ensure complete mixing. Wear rubber gloves when mixing manure samples with your hands.
5. Collect about one quart of manure from the composite sample and place in an appropriate container.
6. Freeze the sample if you are mailing it. Use rapid delivery to ensure that it arrives at the laboratory within 24–48 hours. You can refrigerate the sample if you are delivering it directly to the lab.

Laboratories report results on an as-received or a dry weight basis. As-received results usually are reported in units of lb/ton, while dry weight results usually are reported in percent, ppm, or mg/kg. The "as-received" results, are easily used to determining application rates. Dry-weight results can be used to compare analyses over time and from different manure sources.

To convert manure analyses reported on a dry-weight basis (in percent) to an as-received basis (in lb/wet ton), multiply by 20 to convert the dry weight percent to lb/ton; then multiply by the decimal equivalent (23%/100) of the solids content.

Example: For beef manure at 23% solids and 2.4% nitrogen (N) on a dry weight basis:

Step 1. 2.4% x 20 = 48 lb N/ton dry weight

Step 2. 48 lb N/ton dry weight x 0.23 = 11 lb N/ton as-is.

Analyses typically include total nitrogen, ammonium nitrogen (NH_4^+–N), total phosphorus, total potassium, electrical conductivity, and solids (dry matter). If the manure is old or has been composted you may also want to test for nitrate–N. Total carbon (C) and pH are also useful measurements. Total C can be used to determine the C:N ratio and predict whether or not manure addition is likely to cause nitrogen immobilization. Manure with a C:N ratio greater than 25 is likely to 'tie up' or immobilize nitrogen when you apply it to the soil and stimulate a flush of growth by bacteria and fungi. Bedded manures typically have higher C:N ratios.

MANURE APPLICATION RATES

When application rates of manure are based on providing adequate nitrogen for crop growth, added phosphorus and potassium levels will often exceed crop need, so manure should not ve the sole N source in an organic system. Excess levels of soil P can increase the amount of P in runoff, increasing the risk of surface water pollution. Many crops can handle high levels of K, but livestock can be harmed by nutrient imbalances if they consume a diet of forages with high K levels. Annual P-based manure or compost application is the most effective method of application when soil P buildup is a concern. Phosphorus-based application rates improve water quality, but reduce the amount of manure applied per area and so increase the land base needed for manure application. Where P buildup is a concern, legumes should be included in the rotation to provide additional nitrogen.

Typically manure is applied before the most N-demanding crop in the rotation and the amount of N likely to be plant available during the year of application is estimated. Nitrogen availability from manure varies greatly, depending on the type of animal, type and amount of bedding, and age and storage of manure. Manure contains nitrogen in the organic and ammonium forms. The organic form releases N slowly, while ammonium–N is immediately available for crop growth.

Table. Manure N availability in the first year after application.

Manure type	Total N content (%)	% available N
Broiler litter	4–6	40–70
Laying hen	4–6	40–70
Sheep	2.5–4	25–50
Rabbit	2.5–3.5	20–40
Beef	2–3	20–40
Dry Stack	1.2–2.5	20–40
Separated Solids	1–2	0–20
Horse	0.8–1.6	0–20

Solid manures contain most of their nitrogen in the organic form, but poultry manure contains substantial ammonium–N and so should not be surface applied to avoid loss of ammonia gas. Poultry and other manures that contain a large proportion of ammonium–N should be tilled into the soil the same day they are spread. Ammonia loss is greater in warm, dry, and breezy conditions where soil pH is high and is reduced in cool, wet weather. The N availability numbers are approximate ranges for each type of manure. Use the lower part of the ranges if ammonia losses are likely, the manure contains large amounts of bedding, or if the measured N content is lower than typical values.

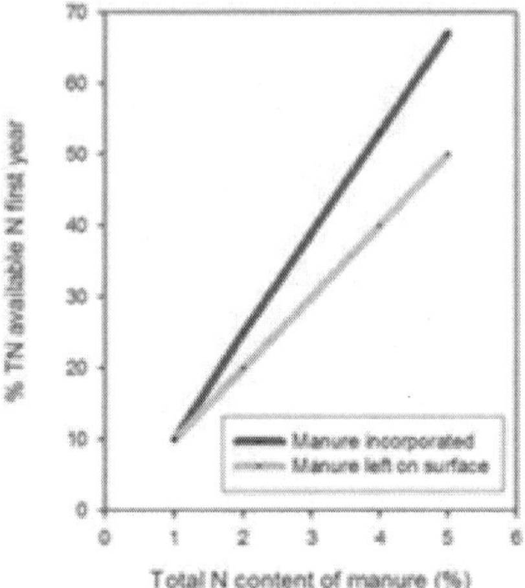

Fig. This graph can be used to predict N release based on total N content during the first year after manure application.

Use the upper range if the manure contains little bedding or if the measured N content is high. Expect first year N tie-up from manures containing less than 1% N. Horse manure or other manures with lots of woody bedding may temporarily tie up nitrogen rather than supply nitrogen for crop growth because the wood is still decaying and bacteria that break down the carbon in the wood consume nitrogen. Composting generally reduces the rate of release of manure N by as much as 50% by converting N into more biologically resistant forms.

"Organic Fertilizer Calculators" like the one linked here are extremely useful tools. Calculators and manure test information are a far better way to calculate application rates than using tables to estimate manure nutrient content and availability.

MONITORING SOIL NUTRIENT LEVELS

Repeated applications of manure can increase the pool of slow-release nutrients and so the amount of manure needed to meet crop needs will decline over time. Farmers need to reduce the manure application rate for fields that receive repeated manure applications.

The trends where repeat application of manure to a clay soil at three rates has influenced soil properties. Organic matter levels were only maintained where rates equaled 20 tons or more. At these rates P and K levels were in excess.

To avoid manure-induced imbalances, continually monitor soil fertility, using appropriate soil tests. Use cover crops and lime or other supplementary fertilizers and amendments to ensure soil balance or restrict application levels if needed.

Table. Effect of 11 years of annual manure additions on the properties of a heavy clay soil planted to continuous corn silage in Vermont.

P					
	Original Level	Application Rate (tons/acre/year)			
		0	10	20	30
Soil organic matter (%)	5.2	4.3	4.8	5.2	5.5
CEC (meq/100g)	17.8	15.8	17	17.8	18.9
pH	6.4	6.0	6.2	6.3	6.4
P (ppm)	4	6.0	7.0	14	17
K (ppm)	129	121	159	191	232
Total Pore Space (%)	n.d.	44	45	47	50

You can use basic soil tests to evaluate the soil for sufficiency or excess of other nutrients. A basic soil test includes P, K, calcium (Ca), magnesium (Mg), boron (B), pH, EC, and a lime recommendation. If you have consistently low levels of P and K and reduced crop growth, you can probably increase your manure application rates. If you have excessive levels of P and K, you should decrease or eliminate manure applications.

Soil tests can be timed to evaluate different aspects of nutrient supply. Testing the year after manures are applied to assess increased P and K supply is recommended. How test levels rise for several years due to the slow release nature of P contained in many manures.

Table. The amount of available phosphorus, expressed as fertilizer equivalent P per tonne or m^3, made available in years following manure application increases over time and vaires with the type of manure added.

			Fertilizer equivalent P per tonne		
	Years after application	*n*	Median	Mean	*P*
Cattle FYM	1	39	0.48	0.62	
	2	55	0.75	1.21	*0.006*
	3	23	1.1	1.82	*0.018*
	4+	8	2.2	2.61	0.12
Pig manure	1	11	0.44	0.56	
	2	11	0.93	0.89	*0.04*
Poultry manure	1	5	1.21	2.44	
	2, 3	7	5.23	5.89	*0.04*
Cattle slurry	1	2	-0.37	-0.37	
	2, 3	5	1.63	2.92	*0.05*
Pig slurry	1	3	0.82	0.87	
	2	6	0.97	1.24	0.46
	3	3	1.53	1.13	*-0.84*

P-values are for the mean FEP being the same as at Ct - 1; italics if $P < 0.05$. Standard errors for each mean are about 0.5 kg/t for FYM and raw cake, 0.2 for pig manure. The local digested cakes contained about 5 kg total P per tonne and raw cake 2.5 kg total P per tonne.

Late season sampling (0–12 inches or more) can be used to determine whether there is surplus nitrate-N remaining in the soil in the fall. If you apply too much manure, unused nitrate N will accumulate. When the fall and winter rains come, the nitrate will leach from the soil and become a potential contaminant in groundwater or surface water. Excess N can also harm some crops, delaying fruiting and increasing the risk of disease damage, freeze damage, and wind damage. Take a "report card" sample as you would any other soil sample, collecting soil cores at multiple spots in the field, and combining the cores together into a composite sample. If late season nitrate–N results are greater than 15–20 mg/kg, you are supplying more N than your crop needs and should reduce or avoid manure application. Report-card nitrate-N levels greater than 30 mg/kg are excessive.

Nutrient requirements for specific crops can be found in Cooperative Extension production guides or from soil test recommendations. Crop performance can be used to help fine tune application rates.

FERTILIZER AND LIME MANAGEMENT

Several questions dealt with management of commercial fertilizer and aglime. The percentage of forage land that receives fertilizer is larger in the cold forest soil region of Minnesota, and when farmers use soil tests. It is reasonable to expect that farmers who use soil tests on forage fields probably are more interested in providing adequate nutrients than farmers who do not soil test. In contrast, less forage land is fertilized where manure was topdressed on established alfalfa (probably due to farmers crediting the nutrient input by manure) and on those farms with permanent grass-legume mixtures. This latter characteristic may reflect farms that have soils not well suited to alfalfa.

Only small numbers of farmers provided estimates of fertilizer rate, so the following relationships between farm characteristics and nutrient application rate are tentative. About 15% of respondents apply P before seeding perennial forages. Reported rates range from about 5 to 60 lb P/acre (median = 16 lb P/acre, average = 21 lb P/acre). About one-quarter of farmers said they apply K before seeding forages, at rates ranging from 24 to 207 lb K/acre (median = 100 lb K/acre, average = 107 lb K/acre). Greater amounts of K are incorporated before seeding when farmers get information from fertilizer dealers and from the Extension Service, and as the proportion of fertilized forage land increases. The rate is smaller in the prairie soil region, which has higher native soil K availability than the rest of the state.

Aglime is added before seeding perennial forages on about one-quarter of the farms, with rates ranging from about 0.30 to 8 tons/acre (median = 3.0 tons/acre, average = 2.8 tons/acre). Lowest rates were for a highly soluble, finely ground limestone. More aglime is used on larger farms and when fertilizer rate information is obtained from fertilizer dealers and farm magazines, while less is applied, as expected, to farms in the calcareous (high pH) prairie soil area.

About 16% of the farmers reported that they apply N fertilizer on alfalfa, with rates ranging from about 5 to 60 lb N/acre (median = 18 lb N/acre, average = 21 lb N/acre). Application rates are likely to be higher when farmers depend on farming magazines for fertilizer rate information and when greater proportions of forage land are fertilized on a farm. Topdressed N applications on alfalfa are smaller on farms with larger herds. When inoculated with rhizobia, established alfalfa often does not respond to N fertilizer, so it is curious that N is applied. The reason may be that only combination fertilizers are recommended or available, that much of the alfalfa acreage contains grasses or other plants that respond to N, or that the farmers do not realize that no N is needed. It is highly unlikely that this 'extra' fertilizer N causes environmental problems, because alfalfa will reduce symbiotic N fixation when fertilizer N is available. More Minnesota dairy farmers reported that they apply N to forages other than to alfalfa. These farmers concentrate topdressed N in one or two applications, with most (64%) being applied in late spring (May-June, M-J).

About one-half of the dairy farmers fertilize before seeding alfalfa, while over 85% fertilize established alfalfa fields. Some farmers (about 12%) reported that they apply the same rate of fertilizer to all perennial forage crop land. For those that do not, fertilizer decisions for both alfalfa and other perennial forages are based mainly on soil tests. Estimated nutrient removal, visual appearance of the crop, amount of grass present, and land ownership also play important roles. Fertilizer P is topdressed at rates ranging from about 5 to 70 lb P/acre (median = 20 lb P/acre, average = 24 lb P/acre) on 26% of the farms. Phosphorus rates are higher in the prairie soil region, on farms where a large percentage of forage land is fertilized, and when information on fertilizer rates is obtained from the Extension Service.

Nearly one-half of Minnesota dairy farmers topdress K on alfalfa. Reported rates range from about 16 to 350 lb K/acre (median = 110 lb K/acre, average = 124 lb K/acre). Relatively little variation in applied K could be explained by farm characteristics, but rate tends to increase with the relative importance of manure pack in livestock housing and decreases with the relative importance of daily or frequent hauling. Presumably, farmers have adopted the practice of K application to established alfalfa with a wide variety of dairy operations.

Topdressed S reportedly is applied to alfalfa by about 10% of Minnesota dairy farmers, and about 5% apply topdressed B. Sulfur rates range from about 6 to 70 lb S/acre and B rates range from 1 to 5 lb B/acre.

These non-N fertilizers are applied almost entirely during the growing season and are spread all at once or in a simple split. A significant number of Minnesota dairy farmers (about 45% of respondents who gave information on timing) apply non-N fertilizer in September and October.

SOIL TEST RESULTS

Farmers were asked to include results of a recent soil test, but only 18% provided soil pH, extractable soil P, and exchangeable K. Many farmers indicated that they did not know the results, because they were on file at the fertilizer dealership. Nearly all of the soil tests reportedly were taken from alfalfa fields. Average soil pH was 6.7 for the cold forest soil area, 7.0 for the warm forest soil area, and 7.4 for the prairie soil area. About 17% of these tested fields were lower than pH 6.5, which is the minimum soil pH recommended for alfalfa in Minnesota. Tests may have been taken in preparation for seeding alfalfa, in which case aglime may have been added after receiving the test results. However, farmers should be reminded that soil pH is an important factor in maximizing alfalfa stand establishment, yield, and persistence.

Almost 80% of the reported soil tests were in excess of 20 mg P/kg, above which no additional P is recommended by the University of Minnesota, and several tests exceeded 120 mg P/kg. Higher soil test P levels were associated with manure storage in a pack and with higher percentages of forage land that are fertilized. Lower soil test P concentrations are characteristic of farms in the prairie soil region and occurred more frequently on farms with daily or frequent haul manure systems. The median soil test P level was 18 ppm for the prairie soil area, but 42 ppm for the warm forest soil area and 50 ppm for the cold forest soil area. About one-half of the reported soil test K values were greater than 160 mg K/kg, the level at which no K fertilizer is recommended. Soil test K concentrations were related positively to pack and lagoon manure storage systems, with increasing proportion of forage land fertilized, and with increasing acreage of permanent grass forages. Pack and lagoon manure handling systems tend to conserve K, which is excreted in urine. There were no differences in reported soil test K among the three soil areas of the state.

It appears from these results that many Minnesota dairy farmers apply commercial fertilizer to alfalfa and other forages. To some extent, the results indicate that native soil characteristics are taken into account when making decisions on liming and nutrient addition, but the large number of high P and K soil tests tend to confirm reports that these nutrients are often in excess on contemporary farms.

MANURE MANAGEMENT

Farmers were asked to describe their manure management systems. The two systems ranked as most typical were daily or frequent hauling (63%) and lagoon (and pit) storage of liquid manure (31%). Dairy farmers apparently rely more on daily or frequent manure hauling in the warm forest soil area of the state, as the percent of forage land that is fertilized increases, and as fertilizer dealers serve as more important sources of information on nutrient needs. This latter point also might be interpreted that farmers who rely on daily or frequent hauling are less likely to employ independent consultants than to rely on personnel at the fertilizer distributor. Daily hauling is less likely on farms with more alfalfa acreage.

Reliance on lagoon storage of liquid manure is more likely as herd size and rolling herd average increase, as one might expect due to environmental regulation of animal feeding operations, and is less likely in the warm forest soil region and when information is obtained from farming magazine articles. Other manure handling systems rarely were listed as most typical, but were employed on many farms. Overall, Minnesota dairy farmers reported that manure storage facilities need to be emptied an average of twice a year, with a range of 1 to 26 times per year.

About one-half of the respondents reported that they apply manure on perennial forages. These dairy farmers are more likely to spread manure before seeding perennial forages when the typical manure source is manure pack in livestock facilities. Manure application either before seeding or as a topdressing on established forage stands is more likely when the farms have larger herds and are in the cold forest soil area of state. Farmers are less likely to apply manure to perennial forages when a larger percentage of their forages received commercial fertilizer. This implies that farmers are distributing manure and fertilizer nutrients to the various crops grown on the farm, and substituting one for the other to some extent. Topdressing manure is less likely the more farmers rely on independent consultants for information on recommended fertilizer rates. Topdressing manure on established grasses is less likely at higher rolling herd averages.

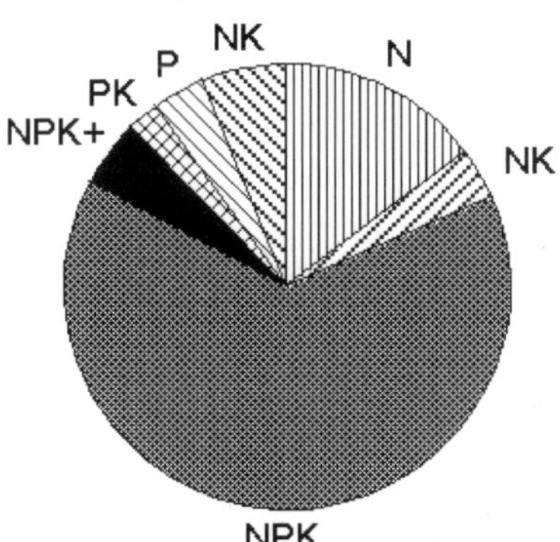

About 6% of all respondents said they routinely apply manure to old forage stands before rotating to the next crop. Farmers should recognize that, although these fields are handy places to apply manure, this practice may be environmentally damaging if N fertilizer rates to the following crop are not reduced based on reasonable N credits for the manure and the previous forage crop.

Only 45% of those who apply manure to forages (at any time) answered the question about whether they reduced subsequent N, P, or K fertilizer additions, and 85% of these replied "yes." About two-thirds of these asserted that they reduced application of all three nutrients, while the remainder reduced application of one or two of these nutrients.

Nearly one-third of the respondents who gave information on manure timing indicated that they topdressed at least a portion of their manure on forages during winter. Spreading manure on frozen soil can lead to serious surface water contamination during snow melt or early spring rains. Farmers should be aware of this potential problem, and select fields for winter spreading to reduce the chance of runoff.

Perennial forages offer a good area to spread manure during the growing season. About one-sixth of respondents spread manure on forages in all months of the year, because they rely on frequent or daily hauling. In contrast, over one-third reported that they spread nearly all the manure used on forages at one time and almost 60% of respondents said they apply manure to forages in two applications.

The most frequently given reasons in support of topdressing manure on established perennial forages included the opportunity to spread manure in summer and to make good use of nutrients. In contrast, at least 20% of the farmers said that lack of time, lack of uniformity in spreading, and increased weed problems were reasons for *not* topdressing manure on all perennial forage fields. Nearly 20% of farmers also replied that they used livestock manure mainly on other crops.

FARMER CONCERNS

About 10% of the farmers indicated that they have concerns about forage nutrition. Of these, nearly one-half were concerned about excess K in dry cow feed because of milk fever, foot and leg problems, and calving problems. Several said that their consultants had made them aware of the problem. Other concerns spanned the range between soil nutrient levels being too low to potential for nitrate poisoning.

INFORMATION SOURCES

When asked what their regular sources of information on recommended fertilizer rates were, over 60% of the farmers reported that they use soil test results as one form of information. More than one-half of the farmers reported that they depend on personnel from fertilizer dealerships for nutrient recommendations. These Minnesota dairy farmers rely more on fertilizer dealers when they have more grass pasture, typically use a daily or frequent haul system for manure, and fertilize a higher proportion of their forage land. However, those who said pasturing their livestock is a typical way of managing manure are less likely to depend on employees of a fertilizer dealership for fertilizer recommendations. Similarly, farmers are more likely to use information from independent crop consultants when they fertilize a larger proportion of their forage land, but also when they have larger herds. Those dairy farmers who spread manure on established forage stands are less likely to use an independent crop consultant for fertilizer recommendations.

Over three-quarters of the farmers rend to rely on information from personnel at fertilizer dealerships to interpred soil test results, and fewer farmers relying on private consultants (27%), Extension (20%), or farming magazine articles (20%) for help in interpreting these tests. Of those farmers who get nutrient management recommendations from fertilizer dealers, 16% also use independent consultants, 13% use Extension, and 20% use farming magazines as other, and sometimes primary, sources of information. Only a small proportion of those who get nutrient management information from independent consultants also use Extension (4%) or farming magazines (5%). Farmers who rely primarily on fertilizer dealers for this information had fewer total cows (median = 49 cows, average = 65 cows) and fewer acres of alfalfa (median = 60 acres, average = 73 acres) than farmers who rely primarily on independent consultants (median = 85 cows, average = 100 cows; median = 120 acres of alfalfa, average = 119 acres of alfalfa). Although Extension does not appear to be a primary source of nutrient management information for many Minnesota dairy farmers, it is part of the mix of information sources for about 14% of them.

5

Organic Manures: Their Nature and Characteristics

ORGANIC MANURES ORGANIC MANURES

Concentrated organic manures have more nutrient con- tents than bulky organic manures. They are oil cakes and meals.

OILCAKES

Oilcakes are very rich in nutrient contents and though insoluble in water, are quick acting organic manures. Their nitrogen becomes quickly available to the plants in about a week or ten days time after application. Mahua cake, however, takes two months to decompose. Oilcakes need to be well powdered before application.

They can be applied few days before sowing or as top dressing (except mahua cake). Oil cakes are more effective in moist soil and wet weather than in dry soil and dry weather.

The use of oilcakes on food grain crops such as wheat and rice is not recommended now on economic grounds. Cakes, especially groundnut and coconut cakes are extensively applied as top dressing in sugarcane crop. Farmers growing betel also use oil-cakes.

MEALS

These are all quick acting manures suitable for all types of soils and crops. This group includes blood meal, fish meal, horn meal, hoof meal and bone meal.

Application of meat meal and blood meal is similar to oilcakes; the latter should be powdered before application. Horns and hooves (slaughter house remnants) are converted into meal by cooking in the bone digester and then dry and powdering them. Most of them are processed in factories and then sent for marketing.

The NPK content of meals are given in Table.

Table. Average Nutrient Content of

Concentrated Organic Manures			
Sl. NoName of Manure	**Percentage composition**		
	Nitrogen	**P_2O_5**	**K_2O**
1. A. Cakes: Non-edible oil cakes			
1 Castor cake	4.3	1.8	1.3
2 Cotton seed cake (undeco- ticated)	3.9	1.8	1.6
3 Karanj cake	3.9	0.9	1.2
4 Mahua cake	2.5	0.8	1.8
5 Safflower cake (undecoticated)	4.9	1.4	1.2
6 Neem cake	5.2	1.0	1.4
B. Edible oil cakes			
1 Coconut cake	3.0	1.9	1.8
2 Cotton seed cake	6.4	2.9	2.2
3 Ground nut cake	7.3	1.5	1.3
4 Linseed cake	4.9	1.4	1.3
5 Jambo cake	4.9	1.6	1.9
6 Niger cake	4.7	1.8	1.3
7 Rapeseed cake	5.2	1.8	1.2
8 Safflower cake (decorticated)	7.9	2.2	1.9
9 Sesamum cake	6.2	2.0	1.2
2. Meals			
1 Blood meal	10-12	1.5	1 0
2 Fish meal	4-10	3.9	0.3-1.5
3 Meat meal	5-10	2.5	0.5
4 Horn and hoof meal	13	—	—
5 Raw bone meal	3.4	20-25	—
6 Steamed bone meal	1.2	25-30	—

EFFECTS OF ORGANIC MANURES ON SOIL PROPERTIES

Addition of organic manures results in multi pronged positive effects on the soil.

Effects on Physical Properties

The organic matter supplied by the manures is a vital component of soil. Water holding capacity and nutrient retaining capacity of the soil are increased. It helps in developing granular structure of the soil. This structure provides enough room for both air and water to stay in the soil. Soil, in the presence of organic matter, develops resistance against the erosion. Manures also impart to the soil a capacity to tolerate extremes of temperature and enable the soil to maintain a narrow and desirable range of temperature.

Effects on Chemical Properties

Organic manures, especially green manures, have already established their importance in the reclamation of alkali soils. They can also correct acidity of the soil to a certain extent, if acidity in the soil is due to presence of acid forming minerals. In other words, we can say that organic manures impart buffering capacity to the soil. Organic matter supplied by the manures to the soil keeps the plant nutrients bound on itself and supply to the plant in times of need. Organic matter also counteracts the adverse effect of heavy metals supplied to the soil through pesticides and also neutralizes acidity or alkalinity created by fertilizers.

Effects on Biological Properties

Soil micro-organisms are like the workers in the factory that is soil. They bring about numerous biological transformations in the soil. Organic matter serves as food as well as raw material for these micro-organisms.

PRINCIPLES OF MANURING

In their natural condition all soils not absolutely barren are capable of supporting a certain amount of vegetation, and they continue to do so for an unlimited period, because the whole of the substances extracted from them are again restored, either directly by the decay of the plants, or indirectly by the droppings of the wild animals which have browzed upon them. Under these circumstances, a soil yields what may be called its normal produce, which varies within comparatively narrow limits, according to the nature of the season, temperature, and other climatic conditions.

But the case is completely altered if the crop, in place of being allowed to decay on the soil, is removed from it, for, though the air will continue to afford an undiminished supply of those elements of the food of plants which may be derived from it, the fixed substances, which can only be obtained from the soil, decrease in quantity, and are at length entirely exhausted.

In this way a gradual diminution of the fertility of the soil takes place, until, after the lapse of a period, longer or shorter, according to its natural resources, it will become entirely incapable of maintaining a crop, and fall into absolute infertility unless the substances removed from it are restored from some other source in the form of manure.

When this is done, the fertility of the soil may not only be sustained but greatly increased, and, in point of fact, all cultivated soils, by the use of manure, are made to yield a much larger crop than they can do in their natural condition.

The fundamental principle upon which a manure is employed is that of adding to the soil an abundant supply of the elements removed from it by plants in the condition best fitted for absorption by their roots; but looked at in its broadest point of view, it acts not merely in this way, but also by promoting the decomposition of the already partially disintegrated rocks of which the soil is composed, setting free those substances it already contains, and facilitating their absorption by the plants.

In considering the practical applications of the broad general principle just stated, it might be assumed that a manure ought invariably to contain all the elements of plants in the quantities in which they are removed by the crops, and that when this has been accurately ascertained by analysis, it would only be necessary to use the various substances in the pro this, though a very important, and no doubt in many cases essential condition, is by no means the only matter which requires to be taken into consideration in the economical application of manures. And this becomes sufficiently obvious when the circumstances attending the exhaustion of the soil are minutely examined.

When a soil is cropped during a succession of years with the same plant, and at length becomes incapable of longer maintaining it, the exhaustion is rarely, if ever, due to the simultaneous consumption of all its different constituents, but generally depends upon that of one individual substance, which, from its having originally existed in the soil in comparatively small quantity, is removed in a shorter time than the others. To restore the fertility of a soil in this condition, it is by no means necessary to supply all the different substances required by the plant, for it will suffice to add that which has been entirely removed.

On the other hand, if an ordinary soil be supplied with a manure containing a very small quantity of one of the elements of plant food, along with abundance of all the others, the amount of increase which it yields must obviously be measured, not by those which are abundant, but by that which is deficient; for the crop which grows luxuriantly so long as it obtains a supply of all its constituents, is arrested as effectually by the want of one as of all, as has been proved by the experiments of Prince Salm Horstmar and others; and hence, in order to obtain a good crop, it would be necessary to use the manure in such abundance as to supply a sufficiency of the deficient element for that purpose.

If this course were persevered in for a succession of years, the other substances which would have been used in much more than the quantity required by the crops, must either have been entirely lost or have accumulated in the soil.

In the latter case it is sufficiently obvious that the soil must have been gradually acquiring an amount of resources which must remain dormant until the system of manuring is changed. To render them available, it is only necessary to add to it a quantity of the particular substance in which the manure hitherto employed has been deficient, so as to restore the lost balance, and enable the plant to make use of those which have been stored up within it. The substance so used is called a *special* manure; that containing all the constituents of the crop is a *general* manure.

The distinction of these two classes of manures is very important in a practical point of view, because a special manure is not by itself capable of maintaining the life of plants, but is only a means of bringing into use the natural and acquired resources of the soil.

In place of preventing or retarding its exhaustion, it rather accelerates it by causing the increased crops to consume more abundantly, and within a shorter period of time, those substances which it contains. On the other hand, a general manure prevents or diminishes the consumption of the elements of plant-food contained in the soil, and if added in sufficient abundance, may cause them to accumulate in it, and even enable an almost absolutely barren soil to yield a tolerable crop.

General manures must therefore always be the most important and essential, and no others would be used if it were possible to obtain them of a composition exactly suited to the requirements of the crop to be raised.

Practically, however, this condition cannot be fulfilled, because all the substances available for the purpose, and particularly farm-yard manure, are refuse matters, the exact composition of which is not under our control, and they do not necessarily contain their constituents either in the most suitable proportions, or the most available forms, and consequently when they are used during a succession of years, certain of their constituents may accumulate in the soil, and it is under such circumstances that special manures are both necessary and advantageous.

Several different substances, but more especially farm-yard manure, fulfil in a very remarkable manner the conditions of a general manure, and supply abundantly, not merely the mineral, but also the carbonaceous and nitrogenous matters necessary for building up the organic part of the plant; and hence its use is governed by principles of comparative simplicity, and really resolves itself into determining the best mode of managing it so as effectually to preserve its useful constituents, and, at the same time, to bring them into those forms of combination in which they are most available to the plant.

But the employment of a special manure opens up nice questions as to the relative importance of the different elements of plants which have given rise to much controversy and difference of opinion. In treating of the food of plants, it has been already observed that the fixed or mineral constituents which are contained in their ash, are necessarily derived exclusively from the soil, but that the carbon, hydrogen, nitrogen, and oxygen, of which their organic part is composed, may be obtained either from that source or from the air.

The important distinction which thus exists between these two classes of substances, has given rise to two different views regarding the theory of manures. Basing his views on the presence of the organic elements in the air, Liebig has maintained that it is unnecessary to supply them in the manure, while others, among whom Messrs.

Lawes and Gilbert have taken a prominent position, hold that, as a rule, fertile soils, cultivated in the ordinary manner, contain a sufficient supply of mineral matters for the production of the largest possible crops, but that the quantity of ammonia and nitric acid which the plants are capable of extracting from the air is insufficient, and must be supplemented by manures containing them.

A large number of experiments have been made in support of these views, but the inferences which can be drawn from them are not absolutely conclusive on either side, and it is necessary to consider the matter in a general point of view.

Setting out from the proposition already so frequently referred to, that the plant cannot grow unless it receives a supply of all its elements, it must be obvious that if, to a soil containing a sufficiency of mineral matters to raise a given number of crops, a supply of ammonia be added, its total productive capacity cannot be thus increased; and though it may yield larger crops than it would have done without that substance, this can only be accomplished by a proportionate diminution of their number.

In either case, the same quantity of vegetable matter will be produced, but the time within which it is obtained will be regulated by the supply of ammonia. That substance differs in no respect from any other element of plant-food, and used in this way is to all intents and purposes a special manure, and acts merely by bringing into play those substances which the soil already contains. Its effect may not be apparent until after the lapse of a very long period of time, but it ultimately leads to the exhaustion of the soil.

If, on the other hand, a soil be continuously cropped until it ceases to yield any produce, it is manifest that the exhaustion must in this instance be entirely due to the removal of its available mineral nutriment, because the superincumbent air constantly changed by the winds must continue to afford the same unvarying supply of the organic elements, and the power of supporting vegetation would be restored to it, by adding the necessary inorganic matters.

Hence when a soil, which in its natural condition is capable of yielding a certain amount of vegetable matter, is rendered barren by the removal of the crop, it may be laid down as an incontrovertible position, that its infertility is due to the loss of mineral matters, and that it may be restored to its pristine condition by the use of them, and of them only. But the case is materially altered when we come to consider the course of events in a cultivated soil.

The object of agriculture is to cause the soil, by appropriate treatment, to yield much more than its normal produce, and the question is, how this can be best and most economically effected in practice. According to Liebig, it is attained by adding to the soil a liberal supply of those mineral substances required by the plant, and that it is unnecessary to use any of the organic elements, because they are supplied by the air in sufficient quantity to meet the requirements of the most abundant crops.

Other chemists and vegetable physiologists again hold that though a certain increase may be obtained in this way, a point is soon reached beyond which mineral matters will not cause the plant to absorb more ammonia from the air, although a further increase may be obtained by the addition of nitrogen in that or some other available form. It is admitted on both sides, that all the elements of plant food are equally essential, and the controversy really lies in determining what practically limits the crop producible on any soil.

The point at issue may be put in a clear point of view by considering the course of events on a soil altogether devoid of the elements of plants. If a small quantity of mineral matters be added to such a soil, it immediately becomes capable of supporting a certain amount of vegetation, deriving from the air the organic elements necessary for this purpose, and with every increase of the former, the air will be laid under a larger contribution of the latter, to support the increased growth, and this must proceed until the limit of supply from the atmosphere is reached.

At this point a further supply of mineral matters alone must obviously be incapable of again increasing the crop, and it would thus be absolutely necessary to conjoin them with a proportionate quantity of organic substances. Liebig maintains that this limit is never attained in practice, but that the air affords ammonia and the other organic elements in excess of the requirements of the largest crop, while mineral matters are generally though not invariably present in the soil in insufficient quantity. Messrs.

Lawes and Gilbert, on the other hand, believe that the soil generally contains an excess of mineral matters, and that a manure which is to bring out their full effect must contain ammonia, or some other nitrogenous substance fitted to supplement the deficient supply afforded by the atmosphere. In short, the question at issue is, whether there is or is not a sufficiency of atmospheric food to meet the demands of the largest crop which can practically be produced. An absolutely conclusive reply to this question is by no means easy.

The experiments by which it is to be resolved are complicated by the fact, that all soils capable of supporting anything like a crop, contain not only the mineral, but the organic elements of its food in large and generally in greatly superabundant quantity, and it is impossible satisfactorily to ascertain how much is derived from this source, and how much from the atmosphere.

There are in fact no experiments in which the effects of a purely mineral soil have been ascertained. The important and carefully performed researches of Messrs. Lawes and Gilbert were made upon a soil which had been long under cultivation, and contained decaying vegetable matters in sufficient abundance to supply nitrogen to many successive crops, and it would be most unreasonable to assert that the produce they did obtain by means of mineral manures, drew the whole of its nitrogen from the air. On the contrary, it may be fairly assumed that the soil did yield a certain quantity of its nitrogenous compounds, but to what extent this occurs, it is impossible to determine.

This difficulty is encountered more or less in all the other experiments, and precludes absolute conclusions. The same fallacy also besets the arguments of Liebig when he holds that the crop, increased by means of mineral manures alone, must derive the whole of the additional quantity of nitrogen which it contains from the air.

So far from this being the case, it is just as likely that the mineral matters should cause the plants to take it from the soil, if it is there, as from the atmosphere. Taking a general view of the whole question, it is evident that a certain amount of vegetation may always be produced by means of mineral manures, and the quantity obtained is generally much beyond the normal produce of the soil.

But it is still open to doubt whether the largest possible crop can be thus obtained, although the balance of evidence is against it, and in favour of the addition of ammonia, and other nitrogenous and organic substances, to the soil.

In actual practice manures containing nitrogen are more important, and more extensively applied than any others, and the quantity of that element thus used is very much larger than is generally supposed. Twenty tons of farm-yard manure, a quantity commonly applied, and often exceeded on well cultivated land, contain a sufficiency of organic matters to yield about 2-1/2 cwt. of nitrogen.

A complete rotation, according to the six-course shift, contains almost exactly the same quantity of nitrogen, when we assume average crops throughout the whole, and it is thus made up.

The supply is therefore quite sufficient for the requirements of the crop; and when it is borne in mind that a considerable quantity of ammonia and nitric acid is annually carried down by the rain, and that during a long rotation other substances are very generally used in addition to farm-yard manure, it is obvious that the crop need not depend to any extent upon what it derives from the air. What is true of the nitrogenous matters applies with still greater force to the mineral constituents of the manure.

Twenty tons of farm-yard manure contain 32 cwt. of mineral matters, while the average crops of a six course-shift contain only 1088 lbs., or less than one-third of this quantity. It is obvious, therefore, that in well manured land there must be a gradual increase of all the constituents of plants, but that of the mineral matters is relatively much greater than that of the nitrogenous.

If therefore from any cause the crop produced on a soil to which farm-yard manure had been applied were greatly to exceed the average, the amount of produce, so far as the soil is concerned, would be limited not by deficiency of mineral, but of nitrogenous food. Hence also when farm-yard manure is liberally applied, there is a gradual accumulation of valuable matters, and a progressive improvement of the productive capacity of the soil.

It is far otherwise, however, if a special manure is employed, because in that case the crop is thrown upon the resources of the soil itself for all its constituents except those contained in the substance employed, and by persisting in its exclusive use exhaustion is the inevitable result. It would be wrong, however, to infer from this, that special manures are to be avoided. On the contrary, great benefits are derived from their judicious employment, and the circumstances under which they are admissible may be readily gathered from what has already been said. They are agents which bring into useful activity the dormant resources of the soil, they restore the proper balance between its different constituents, and supply the excessive demand of some particular elements.

Thus, for instance, in a soil containing an abundant supply of mineral matters, a salt of ammonia or nitric acid increases the crop, by promoting the absorption of the substances already present. So likewise a soil on which young cattle and milch cows have been long pastured has its fertility restored by phosphate of lime, because that substance is removed in the bones and milk in relatively much larger proportion than any others.

The choice of a special manure is necessarily dependent on a great variety of circumstances, and is governed partly by the nature of the soil, and partly by that of the crop. It is obvious that cases may occur in which any individual element of the plant may be deficient, and ought to be supplied, but experience has shown that, as a rule, nitrogen and phosphoric acid are the substances which it is most necessary to furnish in this way, and which in all but exceptional cases produce a marked effect on the crop.

The other substances, such as potash, soda, magnesia, etc., occasionally act beneficially, but the results obtained from them are very uncertain, and frequently entirely negative. It has been commonly asserted that phosphates are specially adapted to root crops, and ammonia or nitrates to the cereals, and this statement is so far true, that the former are used with advantage on the turnip, while the latter act with great benefit on grain crops and more especially on oats and barley.

The effect of the latter, however, is more or less apparent in all crops and on all soils, because it promotes the assimilation of the mineral matters already present. But its peculiar importance lies in the power which it has of promoting the rapid development of the young plant, causing it to send its roots out into the soil, and to spread its leaves into the air, thus enabling it to take from those two sources, abundance of the useful substances existing in them.

But it ought to be distinctly understood, that the statement that particular manures are specially suited to particular crops must be assumed with some reservation, because everything depends upon the nature of the food contained in the soil. It is well known that there are many soils in which ammonia acts more favourably on the turnip than phosphates, and *vice versa*, and the difference is often due to the previous treatment. In many cases in which ammonia when first used proved most beneficial, it now begins to lose its effect, and the reason no doubt is, that by its means the phosphates existing in these soils have been reduced in amount, while the ammonia has accumulated, so that a change in the system of manuring becomes necessary.

A general manure may be used year after year in a perfectly routine manner, but where a special manure is employed, the importance of watching its effects, and altering it as circumstances indicate, cannot be over-estimated. The length of time during which special manures have been extensively used has not been sufficient to bring this prominently before the agriculturist, but its importance must sooner or later force itself upon him, and he will then see the necessity for studying the succession of manures as well as that of crops.

Hitherto we have considered a manure merely as a source from which plants derive their food, but it exercises a scarcely less important action on the chemical and physical properties of the soil. Farm-yard manure, which, as we shall afterwards see, contains a large amount of decomposing vegetable and animal matters, yields a supply of carbonic acid, which operates on the mineral constituents, promotes their further disintegration, and thus liberates their useful elements.

It affects also their physical properties, for it diminishes the tenacity of heavy clays; each straw as it decomposes forming a channel through which the roots of plants, air, and moisture can penetrate more readily than through the stiff clay itself.

On the other hand, it diminishes the porosity of light sandy soils, causes them to retain moisture, and generally makes their texture more suitable to the plant. Special manures probably act to some extent chemically on the soil, but the nature of the changes they produce is as yet imperfectly understood. Superphosphates which are highly acid in all probability act powerfully on the mineral substances, and common salt, which, though of little importance to the plant, occasionally produces very striking effects, appears to exercise some decomposing action on the soil. It is difficult, however, to trace the mode in which they operate on a substance of such complexity as the soil.

Lime, as we shall afterwards see, acts by promoting the decomposition of the vegetable matters on the soil, and possibly some other substances may have a similar effect. In the application of manures to the soil there are several circumstances which must be taken into consideration. It is generally stated that they ought to be distributed as uniformly as possible, but this is not always necessary nor even advisable, and certainly is not acted on in practice.

Much must depend upon the nature both of crop and soil. When the former throws out long and widely penetrating roots, the more uniformly the manure is distributed the better; but if the rootlets are short, it is clearly more advisable that it should be deposited at no great distance from the seed. Practically this is observed in the case of the potato and turnip, which are short rooted, and where the manure is generally deposited close to the seed.

But this course is never adopted with the long rooted cereals, the manure being usually applied to the previous crop, so that the repeated ploughings to which the soil is subjected in the interval may distribute what remains as widely and uniformly as possible.

In soils which are either excessively tenacious or light, the accumulation of the manure close to the plants has also the effect of producing an artificial soil in their immediate neighbourhood, containing abundance of plant-food, and having physical properties better fitted for the support of the plant.

On the other hand, when a special manure is used alone, and with the view of promoting the assimilation of substances already existing in the soil, the more uniform its distribution the better, because it is essential that the roots which penetrate through it should find at every point they reach not only the original soil constituents, but also the substances used to supplement their deficiencies.

BENEFITS OF ORGANIC FERTILIZERS

ORGANIC CULTIVATION, ORGANIC FERTILIZERS

Different types of fertilizers, such as chemical fertilizers, organic fertilizers, and natural fertilizers are available on the market. The type of fertilizer you use has a large impact on the quality of your product. Farmers all over the world use chemical fertilizers, but many are now shifting to organic fertilizers due to the apparent benefits of the latter.

Organic fertilizers are carbon-based compounds that increase the productivity and growth quality of plants. They have various benefits over chemical fertilizers, which include the following:

Non-toxic Food: Use of these organic fertilizers ensures that the food items produced are free of harmful chemicals. As a result, the end consumers who eat these organic products are less prone to diseases such as cancer, strokes, and skin disorders, as compared to those who consume food items produced using chemical fertilizers.

On-Farm Production: The majority of organic fertilizers can be prepared locally or on the farm itself. Hence, the cost of these fertilizers is much lower than the cost of chemical fertilizers.

Low Capital Investment: In addition to the on-farm production possibilities of organic fertilizers, organic fertilizers help in maintaining the soil structure and increasing its nutrient-holding capacity. Therefore, a farmer who has practiced organic farming for many years will require far less fertilizer, because his soil is already rich in essential nutrients.

Fertility of Soil: Organic fertilizers ensure that the farms remain fertile for hundreds of years. Land located at the site of ancient civilizations, such as India and China, are still fertile, even though agriculture has been practiced there for thousands of years. The fertility is maintained because organic fertilizers were always used in the past. However, with the increased use of chemical fertilizers today, land is rapidly becoming infertile, forcing many farmers to further increase their use of chemical fertilizers or even leave the farming industry entirely.

Safe Environment: Organic fertilizers are easily bio-degradable and do not cause environmental pollution. On the other hand, chemical fertilizers contaminate both the land and water, which is a major cause of diseases for human beings and is the force behind the extinction of a number of plant, animal, and insect species.

Employment: We all know that chemical fertilizers are made in large plants that are automated and have an annual capacity of millions of tons. Organic fertilizers, on the other hand, are prepared locally and on a much smaller scale. As a result, the production of organic fertilizers leads to employment, especially in rural areas where employment opportunities can sometimes be bleak.

Why Do You Need Fertilizers?

Why do you need fertilizers for gardening or farming? Do plants not grow in the wild without any fertilizers? The answer to this question lies in the fact that you only want specific species of plants to grow in a farm or a garden.

Soil, on farms as well as the wild, has nutrients in it. However, the proportion of minerals is different in different places. Within a field, the nutrient content is similarly not uniform. In the wild, plants that are most suitable to the soil and other conditions grow and you don't have any control over their growth. However, since you want to grow only your favored variety of plants in your garden or farm, you need to add nutrients according to the needs of that variety and availability in your soil.

Moreover, in a partially covered piece of land, such as a farm, the soil nutrients are taken up by the plants, but are not replenished when the plants or their leaves die, because you remove them as food, herbs, etc.. Hence you require fertilizers to enrich the nutrient-deficient soil and create artificial conditions for higher productivity of the plants.

What are these nutrients required by plants? The most commonly required plant nutrients are nitrogen (N), phosphorus (P) and potassium (K). Almost all fertilizers are categorized according to their Nitrogen-Phosphorus-Potassium or N-P-K value. Nitrogen is required for the growth of vegetative parts such as the stems and the leaves, while your plants will have healthy roots if they get a sufficient amount of phosphorus. Phosphorus is also required for good flowers and fruits. Potassium makes the plant healthy by facilitating the circulation of nutrients within the plant.

In addition to N-P-K, plants also require other nutrients, such as calcium and magnesium. Since these are required in small quantities, you need not add them separately unless in exceptional cases, if your soil is totally devoid of these minerals or the crop you wish to grow requires them in large quantities.

How to Apply Fertilizers?

Farmers have the conception that adding large quantities of fertilizers in their farms will only be beneficial to their plants. However, according to The Royal Horticultural Society, since fertilizers are substances that are rich in nutrients, they are meant to be added in small quantities. The society suggests the following points for fertilizer application:

Soil Analysis: A thorough analysis of the soil should be conducted before planting the crops. You should first find out the existing nutrient content of your soil, find out the nutrient requirement of the plants you want to grow, and add fertilizers based on the nutrient deficiency between the two. A typical soil analysis test will determine the soil texture, organic matter content, and pH. It will also give you the content of different minerals such as phosphorus, magnesium, and potassium. Some soil test labs also determine the micro-organic activity in your soil.

Time of Application: The time when you apply the fertilizers is also crucial. In cold climates, fertilizers should not be applied during winter and autumn, as they will favor young growth during the winter. Since the weather is harsh during this time, the plants won't be able to survive for long. Therefore, fertilizers should only be applied during the spring. In fact, the time for adding the fertilizers and preparation of the crop should be quite close.

Quantity of Fertilizers: Society suggests that if a farmer is in doubt about the quantity of fertilizer to be added, he should add less fertilizer to be on the safe side.

Organic Fertilizers: It is also suggested that farmers prefer organic fertilizers to chemical fertilizers.

Global awareness for the hazards of long-term chemical fertilizer use is growing. Because of this, more and more farmers all over the world are shifting to organic fertilizers. The agricultural market has also recognized this trend, and has recently employed a full-blown campaign to promote organic and natural fertilizers. Among the benefits of using organic fertilizers are non-toxic food, lower cost, better soil fertility, and of course, a safer environment.

What is Compost?

One of the best features of organic gardening is the ability to locally produce the fertilizers, pesticides, and other organic ingredients. And the easiest among them is compost. You can prepare compost almost anywhere, even in your kitchen garden!

Agricultural scientists all over the world agree that compost is an excellent source of organic matter for garden plants. Compost provides air, water, organic matter, and microorganisms to your plants, thus enhancing their growth. It also maintains a healthy atmosphere for the soil and hence keeps insects, plant diseases, and weeds away.

What are the Benefits of Compost?

So what is so special about compost? Why not use organic matter directly? The biggest advantage of using compost is that the organic matter in compost is partially decayed, so its volume is much lower. Furthermore, the microorganism activity has started in the compost already, so the concentration of these microorganisms is very high. This makes the compost a concentrated and easy to absorb source of organic matter for the growing plants.

Where can you buy compost? You can purchase compost from the market, as well as online stores. Compost for Sale is a directory of some companies that sell compost in the UK, US, and India.

You can also prepare compost in your own garden. Composting on a small scale can be done through the following three techniques:

Fast Composting: Fast composting is a composting technique used by many biodynamic farmers. A pit of 1m height, 1m width and 1m length is prepared. The length of the pit can be increased according to the space available in your garden and your compost requirement. The pit is filled in with a thin layers of leaves, manure, and straws. Water is added after adding each layer. You can also add some soil between these layers. The pile should be turned regularly to keep it aerated. Compost will be ready within six to eight weeks.

Slow Composting: Slow composting, as the name suggests, is a very slow process and it takes a number of months for compost formation through this method. Organic matter that is rich in carbon (brown organic matter such as saw dust) is the main ingredient. Fill half of the composting pit with carbon rich organic matter, and every day, add kitchen waste or vegetable peelings to this mixture. You can also add soil along with the kitchen waste occasionally. The pit's contents will gradually decompose and you can start harvesting the compost from the bottom of the pit.

Worm Composting: Worm composting, also known as vermiculture, can be carried out in a bin or in a trench. Fill the bin or trench with soil and organic matter (kitchen waste or vegetable peelings). Then, release red worms into the bin. Ensure that the bin is always moist, but don't add too much water. Adding water in large quantities will fill the pores built by the worms with water and disturb their colony. You can also make a hole in the bin for purging out any excess water. Some researchers also suggest using earth worms instead of red worms, since earth worms are more effective in converting organic matter into manure. The compost will be ready in a few weeks. While harvesting the compost, take care that you don't hurt the worms.

QUALITY OF ORGANIC FERTILIZERS

When you prepare compost or organic fertilizers, you can add a variety of organic matter to make the compost. The quality or nutrient value of the compost depends on what organic matter you add to it. Also, if you are purchasing organic fertilizers, you should try to know the nutrient content of the organic fertilizers before purchasing them. Just like chemical fertilizers, organic fertilizers also have an N-P-K (nitrogen potassium and phosphorus) value.

Organic Matter	% Nitrogen	% Phosphorus	% Potassium	Availability of Nutrients
Alfa Alfa Hay	2-3	0.5-1	1-2	Medium
Bone Meal	1	11	0	Slow
Cottonseed Meal	6	3	1	Slow
Compost	1.5	0.5	1	Slow
Dried Blood	12	1.5	0.5	Rapid
Feather Meal	12	0	0	Medium
Fish Meal	10	4	0	Slow
Grass Clippings	1-2	0-0.5	1-2	Medium
Horn Meal	12-14	1.5-2	0	Medium
Kelp	1	0.5	9	Rapid
Leaves	1	0-0.5	0-0.5	Slow
Legumes	2-4	0-0.5	2-3	Medium
Cow Manure	0.25	0.15	0.25	Medium
Horse Manure	0.3	0.15	0.5	Medium
Sheep Manure	0.6	0.33	0.75	Medium
Swine Manure	0.3	0.3	0.3	Medium
Pine Needles Manure	0.5	0	1	Slow
Poultry Manure	2	2	1	Rapid
Saw Dust Manure	0-1	0-0.5	0-1	Slow
Sewage Manure	2-6	1-4	0-1	Moderate
Seaweed Manure	1	2	5	Rapid
Straw Manure	0-0.5	0-0.5	1	Slow
Wood Ashes	0	1-2	3-7	Rapid

The NPK value for organic fertilizers depends on the organic matter used for preparing them. Organic matter such as dried blood and fish meal are rich in nitrogen. Fish meal is also rich in phosphorus. Kelp, on the other hand, is rich in potassium. The ability of different types of organic matter to release nutrients is also not the same. Therefore, the availability of nutrients is also an important criterion when selecting the type of organic fertilizer.

The Cornell Cooperative Extension and The Utah State University Extension have provided the nutrient value of different organic matter in their guides Fertilizing Garden Soils and Selecting and Using Organic Fertilizers, respectively. Some of these values are tabulated below:

NUTRIENT COMPOSITION OF MANURE AND COMPOST

Many different types of manure are available for crop production. It is assumed that most vegetable growers will be using solid manure with or without bedding. Similar principles will apply to the use of liquid manures. The nutrient content of manures varies with animal, bedding, storage, and processing. The approximate nutrient composition of various solid manures, including some composted manures. While this table provides a general analysis of manure or compost nutrient content, it is strongly recommended that if routine applications are made for crop production the specific manure being used should be tested by a laboratory for moisture and nutrient content. Nutrient analysis should include: total nitrogen (N), ammonium-N, phosphate (P_2O_5), and potash (K_2O). Accurate manure or compost analysis requires that a representative sample be submitted; so several subsamples should be collected and composited to make up the sample. If manure or compost is being purchased, request a nutrient analysis from the seller for N, P_2O_5, and K_2O content.

Fresh vs. composted manure. Fresh, non-composted manure will generally have a higher N content than composted manure. However, the use of composted manure will contribute more to the organic matter content of the soil. Fresh manure is high in soluble forms of N, which can lead to salt build-up and leaching losses if over applied. Fresh manure may contain high amounts of viable weed seeds, which can lead to weed problems. In addition, various pathogens such as *E. coli* may be present in fresh manure and can cause illness to individuals eating fresh produce unless proper precautions are taken. Apply and incorporate raw manure in fields where crops are intended for human consumption at least three months before the crop will be harvested. Allow four months between application and harvest of root and leaf crops that come in contact with the soil. Do not surface apply raw manure under orchard trees where fallen fruit will be harvested.

Heat generated during the composting process will kill most weed seeds and pathogens, provided temperatures are maintained at or above 131°F for

15 days or more (and the compost is turned so that all material is exposed to this temperature for a minimum of 3 days). The microbially mediated composting process will lower the amount of soluble N forms by stabilizing the N in larger organic, humus-like compounds. A disadvantage of composting is that some of the ammonia-N will be lost as a gas. Compost alone also may not be able to supply adequate available nutrients, particularly N, during rapid growth phases of crops with high nutrient demands. Composted manure is usually more expensive than fresh or partially aged manure.

Heat-dried manure/compost. Drying manure or compost to low moisture content reduces their volume and weight, which lowers transportation costs, but it also requires energy inputs. Dried products can be easier to handle and apply uniformly to fields, especially those that have been processed into pellets. Heat drying also reduces pathogens if temperatures exceed 150 to 175°F for at least one hour and water content is reduced to 10 to 12% or less. The significant energy costs to heat-dry manure or compost at high temperatures are in contrast to the self-heating generated by microbial respiration during the composting process. Heat-dried composts vary widely in the degree to which they are composted before drying. Many are only partially composted and have higher amounts of soluble (inorganic) N forms than mature, stable compost. This readily available N gives these products some characteristics that are similar to soluble N fertilizers, such as ammonium nitrate. Heat drying of manure and immature compost may increase volatilization of ammonia-N and reduce the total N content of the finished product. In addition, composted or partially composted material that is dried at high temperature rather than going through a curing phase at ambient temperatures is not as biologically active as mature compost. The disease suppressive properties of some composts depends upon recolonization of the compost by disease suppressing organisms during the curing phase.

NUTRIENT AVAILABILITY FROM MANURE AND COMPOST

The analysis of manure or compost provides total nutrient content, but availability of the nutrients for plant growth will depend on their breakdown and release from the organic components. Generally, 70 to 80% of the phosphorus (P) and 80 to 90% of the potassium (K) will be available from manure the first year after application. Numbers from a table or from an analysis report should to be multiplied by these factors to obtain the amount of P_2O_5 and K_2O available to crops from a manure or compost application.

Calculating N availability is more complex than determining P and K availability. Most of the N in manure is in the organic form and essentially all of the N in compost is organic. Organic N is unavailable for uptake until microorganisms degrade the organic compounds that contain it. A smaller fraction of the N in manure is in the ammonium/ammonia or inorganic form.

The ammonium-N form is a readily available fraction. Other inorganic forms such as nitrate and nitrite can also exist, but their quantities are usually very low. Estimated levels of ammonium-N and total N in fresh and composted manure.

When applied to soil, manure, compost, and other organic amendments undergo microbial transformations that release plant-available N over time. Volatilization, denitrification, and leaching result in N losses from the soil that reduce the amount of N that can be used by crops.

Table. Approximate Nutrient Composition of Various Types of Animal Manure and Compost (all values are on a fresh weight basis).

Manure Type	Dry Matter	Ammonium-N	Total Na	P2O5	K_2O
	%	---------- lb/ton ----------			
Swine, no bedding	18	6	10	9	8
Swine, with bedding	18	5	6	7	7
Beef, no bedding	52	7	21	14	23
Beef, with bedding	50	8	21	18	26
Dairy, no bedding	18	4	9	4	10
Dairy, with bedding	21	5	9	4	10
Sheep, no bedding	28	5	18	11	26
Sheep, with bedding	28	5	14	9	25
Poultry, no litter	45	26	33	48	34
Poultry, with litter	75	36	56	45	34
Turkey, no litter	22	17	27	20	17
Turkey, with litter	29	13	20	16	13
Horse, with bedding	46	4	14	4	14
Poultry compost	45	1	17	39	23
Dairy compost	45	<1	12	12	26
Mixed compost: Dairy/Swine/Poultry	43	<1	11	11	10

Note:

[a]Total N = Ammonium-N plus organic N

N availability from manure the first growing season after application. The actual amount available is dependent on manure type, bedding, and whether the manure has been composted. Usually 25 to 50% of the organic-N in fresh manure is available the first year. In addition to the organic fraction, N availability from manure also has to take into account the amount of ammonium-N present. This form of N is readily available for plant uptake, but is prone to losses as ammonia if not incorporated within 12 hours after application. Assuming direct manure incorporation after application, 45 to 75% of the total N (organic-N + ammonium-N) is available the first year. Note that for composted manure, the percentage of the organic N available in the first year following application is much lower than it is for fresh manure. Because there is very little ammonium-N in composted manure, the organic N fraction is basically the same as the total N fraction.

Bedding or litter will usually decrease nutrient content by dilution. If materials high in carbon (C) like straw or wood shavings are used as bedding, N availability may be reduced by the larger C/N ratio of the product. High C relative to N will lead to a tie-up of N, potentially causing N deficiency in the crop. A C/N ratio of 25/1 or greater will lead to N tie-up in the soil. A C/N ratio of less than 25/1 will release N to the crop. The C/N ratio is also an important consideration in the use of various composts, as well as a controlling factor in the composting process itself.

Table. Estimated Organic N Availability (Km) from Manure and Composted Manure the First Season After Application.

Manure Type Â	Organic N *(% available)*
Swine, fresh	50
Beef, no bedding	35
Beef, with bedding	25
Dairy, no bedding	35
Dairy, with bedding	25
Sheep, solid	25
Poultry, no litter	50
Poultry, with litter	45
Horse, with bedding	20
Composted poultry	30
Composted dairy	14

MANURE AND COMPOST APPLICATION

Some of the N in fresh manure will be lost to the atmosphere during application in the form of ammonia gas. The higher the ammonium-N fraction is in manure, the more prone it is to ammonia volatilization. Manure should be incorporated within 12 hours of application to avoid excessive ammonia losses. Unincorporated manure will supply the organic N fraction and at most 20% of the ammonium-N fraction. Incorporation of composted manure is not as critical, because the N is stabilized in organic compounds with little free ammonium present. However, in order to obtain full benefit from compost, incorporation is recommended whenever possible. Manure and compost are often high in soluble salts, so to avoid salt injury seeding operations should take place about 3 to 4 weeks after application.

RESIDUAL NUTRIENTS IN SOIL FROM MANURE AND COMPOST APPLICATION

The residual effects of the manure and compost are important. Some benefit will be obtained in the second and third years following application.

When manure and compost are used to fertilize crops, soil organic matter will increase over time and subsequent rates of application can generally be reduced because of increased nutrient cycling. Continuous use of manure or compost can lead to high levels of residual N, P, and other nutrients, which can potentially be transported to lakes and streams in runoff or leach and pollute the groundwater. Taking into account residual release of N in subsequent years should help to avoid excessive applications. General rules of thumb for N are that organic N released during the second and third cropping years after initial application will be 50% and 25%, respectively, of that mineralized during the first cropping season. Remember that some manures and composts contain high levels of P, so if organic nutrient sources are regularly applied at rates to meet crop N demands, the amount of P in the soil can build up to excessively high levels. Use of soil tests, plant tissue tests, and monitoring of crop growth will help in determining the amount of residual N and other nutrients in the soil and the need for further applications.

ORGANIC MANURES AND BIOFERTILIZERS

The use of organic manures (farmyard manure, compost, green manure, etc.) is the oldest and most widely practised means of nutrient replenishment in India. Prior to the 1950s, organic manures were almost the only sources of soil and plant nutrition. Owing to a high animal population, farmyard manure is the most common of the organic manures. Cattle account for 90 percent of total manure production. The proportion of cattle manure available for fertilizing purposes decreased from 70 percent in the early 1970s to 30 percent in the early 1990s. The use of farmyard manure is about 2 tonnes/ha, which is much below the desired rate of 10 tonnes/ha.

At the present production level, the estimated annual production of crop residues is about 300 million tonnes. As two-thirds of all crop residues are used as animal feed, only one-third is available for direct recycling (compost making), which can add 2.5 million tonnes/year. The production of urban compost has been fluctuating around 6–7 million tonnes and the area under green manuring is about 7 million/ha.

Unlike fertilizers, the use of organic material has not increased much in the last two to three decades. The estimated annual available nutrient (NPK) contribution through organic sources is about 5 million tonnes, which could increase to 7.75 million tonnes by 2025. Thus, organic manures have a significant role to play in nutrient supply. In addition to improving soil physico-chemical properties, the supplementary and complementary use of organic manure also improves the efficiency of mineral fertilizer use.

The use of biofertilizers is of relatively recent origin. Biofertilizers consist of N fixers (Rhizobium, Azotobacter, blue green algae, Azolla), phosphate solubilizing bacteria (PSB) and fungi (lmycorrhizae). A contribution of 20–30 kg N/ha has been reported from the use of biofertilizers. There was good growth in biofertilizer production and use in 1990s. At present, biofertilizers use is about 10 000 tonnes. Among biofertilizers, most growth has occurred with phosphate-solubilizing micro-organisms, which account for about 45 percent of total biofertilizer production and use. Biofertilizer production and use is concentrated in Maharashtra, Tamil Nadu, Karnataka, Madhya Pradesh and Gujarat.

Table Growth in biofertilizer production

Year	Capacity	Production	Distribution
		(tonnes)	
1992/93	5 401	2 005	1 600
1995/96	10 680	6 692	6 288
1998/99	16 446	8 010	5 065
2003/04*	20 000	12 000	10 000

Note:

* Estimated

The Government is promoting the concept of the integrated nutrient supply system (INSS), i.e. the combined use of mineral fertilizers, organic manures and biofertilizers. Farmers are also aware of the advantage of INSS in improving soil health and crop productivity. However, the adoption of INSS is limited by the following constraints:

- Increasing trend to use cow manure as a source of fuel in rural areas;
- Increasing use of crop residues as animal feed;
- Extra cost and time required to grow green-manure crops;
- Handling problems with bulky organic manures;
- Problems in timely preparation of the field when agricultural waste and green manure have to be incorporated and their decomposition awaited;
- Poor and inconsistent crop response to biofertilizers.

NATURAL ORGANIC MANURES

FARM YARD MANURE

It is a mixture of cattle dung, urine, litter or bedding material, portion of fodder not consumed by cattle and domestic wastes like ashes etc. collected and dumped into a pit or a heap in the corner of the backyard. It is allowed to remain there and rot till it is taken out and applied to fields. Well rotten Farm Yard Manure contains 0.5.% N., 0.2 % P_2O_5 and 0.5 % K_2O.

COMPOST

Well rotted plant and animal residue is called compost. Composting means rotting of plant and animal remains before applying in fields. The essential requirements of composting are air, moisture, optimum temperature and a small quantity of nitrogen. It is an activity of micro-organisms and same people recommend addition of suitably prepared inoculums to introduce micro-organisms for decomposing the material.

GREEN MANURING

Green manure crops are grown in the field itself either as a pure crop, or as an intercrop with the main crop, and buried in the same field. The most common green manure crops are sannhemp, dhaincha and guar. Tender green-twigs and leaves are collected from wastelands which are spread in the field and incorporated into the soil. Shrubs and trees are also cut and turned into the soil e.g. Shrubs like glyricidia, sesbania, karanj.

BIOFERTILIZERS

The biofertilizers containing biological nitrogen fixing organisms are of utmost importance in agriculture Advantages Of Biofertilizer i) They help in the establishment and growth of crop plants and trees. ii) They enhance biomass production and grain yields by 10-20 percent. iii) They are useful in sustainable agriculture. iv) They are suitable in organic farming. v) They play an important role in Agrotorestry/ Silvi- pastaural system.

TYPES OF BIOFERTILIZERS

Rhizobium

Most widely used biofertilizer is Rhizobium which colonizes the roots of specific legumes to form tumor like growths called root nodules. These nodules act as factories of ammonia production. The Rhizobium – legume association can fix up to 100-300 KG/N. in one crop season.

Azotobacter

Application of azotobacter has been found to increase yield of wheat, rice, maize, pearl-millet and sorghum by 0-30 p.c. over control. Apart from nitrogen this organism is also capable of producing antifungal and antibacterial compounds, hormones.

Azospirillum

Certain micro-organisms like bacteria and blue green algae have the ability to use atmospheric nitrogen and transport this nutrient to the crop plants. Azospirillum is inoculated to maize, barley, oats, sorghum, pearlmillet and forage crops. It increases grain productivity of cereals by 5-20%, of millets by 30% and fodder by over 50%.

Blue-green algae

The utilization of blue green algae as a biofertilizer for rice is very promising. A judicious use of these algae could provide to the country's entire rice acreage as much nitrogen as obtained from 15-17 lakh tones of urea. Algae also helps to reduce soil alkalinity.

Azolla

A small floating water form Azolla is commonly seen in low land fields and in shallow fresh water bodies. These fern harbors a blue-green algae. Anabaena azollae. The Azolla – Anabaena association is a live floating nitrogen factory using energy from photosynthesis to fix atmospheric nitrogen accounting to 100-150 kg N/ ha/ year from about 40 – 60 tones of biomass.

Mycorrhizae

It is the symbiotic association of fungi with roots of vascular plants. It is useful in increasing phosphorus uptake e.g. in fruit crops like citrus, papaya.

VERMI-COMPOST

It is the method of making compost with the use of earthworms, which generally live in soil, eat bio-mass and excrete it in digested form. This compost is generally called vermi-compost or wormi-compost. It is estimated that 1800 worms which is an ideal population for one sq. meter can feed up to 80 tones of humus per day.

6

Potential of Organic Materials and Plant Nutrients

PROVIDING NUTRIENTS FOR CROPS BY SUPPLYING ORGANIC MATERIAL

NITROGEN

The supply of nitrogen (N) in organic farming is usually provided by legumes. Through a symbiosis with nodule bacteria, these plants are capable of fixing atmospheric nitrogen and making it available to plants.

There are other bacteria that can also fix nitrogen (Actinomyceten which are present in dead wood, soil-bound Azotobacter or Beijerinckia bacteria, which live in association with the tropical fodder-grass Paspalum notatum and other Gramineen).

In paddy rice, the bacteria Anabena azollae is used, which forms a symbiosis with the water fern Azolla, and can, under tropical conditions, fix up to 400 kg N/ha and year, and which is very often used as a green manure for rice crops. As already mentioned, most important is the planting of site-appropriate legumes in crop rotation or agroforestry systems (there are over 12,000 types of legumes in the world). In crop rotation systems, at least 20 per cent of the entire cultivated area should be planted with legumes.

PHOSPHOROUS

The phosphate content of the soil vary just as much as the availability of phosphates (P) for the plants (e.g., tropical soil with its high acid, iron and aluminium content has a very high rate of P fixation, thus applied P-fertilizers become unavailable for the plants). On organic farms emphasis is placed upon increasing the availability of the phosphate content of the soil for the plants. *This is achieved by a biological conversion of unsoluble to soluble P-compounds in the soil (enzymes and plant acids):*

- In fertile soils with sufficient organic matter the conversion to soluble Pcompounds is most efficient: Fertile soils have a more intensive growth of roots and subsequently a more developed network of fine roots increasing the interaction between fine roots and phosphate compounds in the soil. Acids released by the fine roots than dissolve the fixed P-compounds and improves the availability for plants.
- Fertile soils with a high organic substance content encourage the growth of VAMycorhizza, a fungus that lives in a symbiosis with plants having a high capacity to dissolve fixed P-compounds.
- Use of plants that are particularly capable to break up fixed P-compounds (e.g., onions in mixed crop systems with cotton, palms and vanilla in agroforestry systems).
- Organic matter (mulching material, compost) increases the availability of phosphates.
- High pH values and poor phosphate availability can be alleviated by applying silicates.
- Adding rock phosphates is still allowed on organic farms. If composting, then the compost can be directly prepared with the rock phosphate. With some crops (e.g., cotton) the seeds are infested with bacteria that can break up phosphate compounds, in order to ensure that the plants have enough supply during the important early stages of growth.

POTASSIUM

Potassium (K) is easily leached out of sandy soils which contain little organic matter (Humus compounds). *The following strategies are important in organic farming to ensure a sufficient supply of K for the plants:*

1. Regular applications of organic matter will improve the absorption of potassium in the upper soil layers, where it can be reached by the plants' roots.
2. Use of deep-rooting plants to mobilise K in lower soil layers.
3. Integration of plants with a high K-uptake in mixed cultivation systems (e.g., bananas on coffee plantations).
4. A permanent mulching layer, especially in the wet tropics in order to reduce leaching of K.
5. In arid regions with soils poor in K, it can be useful to mix pulverised rock containing mineral clay into the compost.

In case of potassium deficiencies showed by soil analysis it is permitted to use certain potassium salts with a low chlorine content (Muriate of potash/ potassium chloride is not allowed). Wood ash from untreated wood is also allowed.

How to Utilise Organic Matter?

One of the main objectives in organic farming is to minimise the loss of nutrients and to keep them recycling within the farm (and/or to use them).

The following criteria should be heeded in this respect: Burning of crop residues (e.g., common practice on sugarcane plantations) is not allowed in organic farming (exceptions may be permitted by the certification body, e.g., in cases of heavily infested crop residues). When crop residues are burnt, important nutrients and energy is lost.

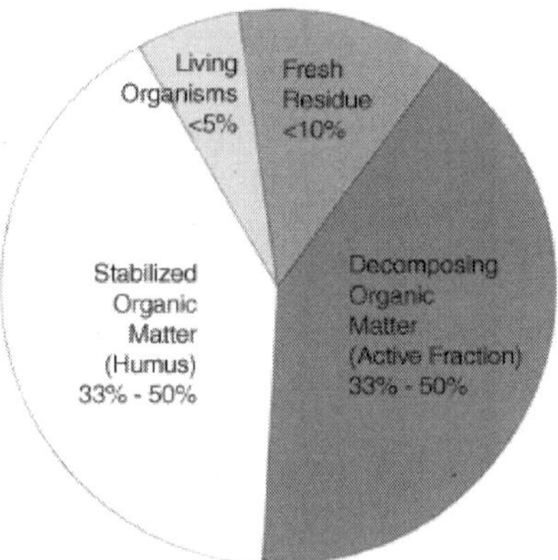

Crop residues should be left on the soil's surface (agroforestry systems) or mulched into the upper layer of the soil (arable farming). Alternatively, they can also be used as fodder or bedding for animals, or used directly as compost material. Organic matter is created during the production and processing of several types of cash crops (coffee, bananas, sugar cane etc.). These residues should be returned into the cycle of nutrients on the farm. This is best achieved by mulching or composting, or by creating bio-gas with the resulting usage of any organic material left over.

Large amounts of pruning material are regularly given, especially in agroforestry systems, which are then also used as fuel. The nutrient rich ashes (minerals like e.g., potassium) should then be fed back into the nutrient cycle via compost.

Farmers will need to decide just how much of the organic material created on their farm should be composted, depending on the site's requirements. Each form of composting will inevitably lead to a certain level of nutrient losses and will increase the working costs.

Note: One objective in organic farming systems is to realise nutrient cycles within a field or farm (or even in a local/ regional) system, closed as much as possible, and to reduce unnecessary losses of nutrients.

With the following examples, describing the situation in "arid climates" and "humid tropics", the above mentioned context can be illustrated as given.

HUMID TROPICS

Growth and decomposition processes are accelerated in these regions due to their high average temperatures with few fluctuations, high rainfall and humidity. Organic material in and on the ground is rapidly degraded and nutrients set free. In humid tropics, covering the ground with a thick layer of organic material and a humus-rich top soil are the most important sources of nutrients and water.

It is therefore essential that this layer of soil is conserved, by leaving it continually covered (if possible) with organic material, by not carrying out any deep or involved tillage, and also, by implementing agroforestry systems to protect against erosion. In addition, planting, and if at all necessary, tillage on sloping sites should only take place along the hang-parallel (contour planting).

Special arrangements for composting are usually unnecessary, as the organic material can be degraded "on site", *i.e.*, where it falls. The aim is not to disturb the upper, fertile, humus-rich layer of soil through agricultural activities, which would lead to the humus being depleted and resulting loss of soil fertility and the additional danger of the soil drying out.

Cattle-raising is not advisable in such areas. Clearing of land by burning organic material is not advisable (slash and burn) because Enormous quantities of valuable nutrients/energy are then lost to the agroeco system.

ARID SITES

Growth conditions for plants are more often than not highly limiting in these areas due to high daily temperatures with large fluctuations, slight rainfall and often very dry winds. The careful use of water as the most important limiting growth factor under these conditions is the key to for the realisation of a site-appropriate farming system.

All possible measures to reduce the losses of water – especially through evaporation and transpiration – should be consequently realised, e.g., sufficient quantities of hedges to act as wind-breaks as well as good a permanent coverage and shading of the soil. In addition, the waterconserving capacity must be increased or at least maintained through the continual application of compost (humus compounds).

Furthermore, it is worth noting that living plants are efficient fresh water savers, and can be effectively integrated into the system as a whole. The irrigation system used must also be capable of minimising water losses and avoiding soil salinity. This system is also characterised by a partial utilisation of the organic matter through animal husbandry and a respective arable fodder cropping system including legumes. Because of the limited biological activity of the soil, the organic matter cannot be degraded in the field (due to a lack of continual water supplies) but as a result of composting.

With the contrary examples given, it becomes evident that in practice, the nearer a farm is to the continuously humid tropics, and away from the changeable climates with their arid periods, the more the agricultural system chosen should adhere to the characteristics of an agroforestry system.

COMPOST

A third primary source of organic matter and nutrients, particularly for smaller plots of land or gardens is compost. Compost is the end product of biological breakdown of organic matter. Composting or breakdown can result from fungal activity at lower temperatures (<90°F) and bacterial activity at higher temperatures (from 120 to 160°F). During composting, carbon from the organic matter is lost as carbon dioxide and heat and water are generated. The resulting material has more concentrated nutrients and can be used as a fertilizer as well as a source of organic matter for soil microorganisms.

The product is more stable, decreases the solubility of nutrients, and avoids the immobilisation of N that can occur with straw or saw dust bedding. The amount of weed seed is also decreased compared to manure and there may be beneficial effects on reducing plant soil born pathogens. The process of composting is also like a cow eating hay. Bacteria are feeding on the carbon, releasing carbon to the atmosphere, but at the same time making nutrients more soluble and more concentrated in the material left behind.

ORGANIC FERTILIZERS AND AMENDMENTS

There are naturally occurring fertilizers or amendments that are acceptable for certified organic production. They can be categorised as either mineral derived, animal derived, or plant derived. Following is a table of several organic fertilizers and a range of rates they can be applied to the garden (in pounds (lbs) per 100 square feet of garden area).

Examples of *mineral derived organic fertilizers*are mined phosphates either in the form of raw rock phosphate, colloidal rock phosphate, or black rock phosphate; a mined potassium silicate based mineral from New Jersey called greensand; lime, which is mined calcium and magnesium carbonate; gypsum, which is calcium sulfate; either potassium sulfate or potassium magnesium sulfate (sol-po-mag) that are also mined, and elemental sulfur.

There is a large, naturally occurring deposit of nitrate of soda, or sodium nitrate in Chile, South America. Sodium nitrate is a very water soluble salt that is usually not considered acceptable.

Examples of *animal derived organic fertilizers* are blood meal (dried blood), bone meal, feather meal, fish meal or fish emulsion, and oyster shell lime.

There are some concerns about the safety of these fertilizers when they are derived from poorly managed animals or animals that were not produced organically. Examples of *plant derived fertilizers* include alfalfa meal, soybean meal, cotton seed meal, sea weed or kelp based materials, and wood ash.

As with some animal derived nutrient sources, there is reservation about using cotton seed meal derived from cotton that is often heavily sprayed with insecticides. Since, genetic modified organisms are not allowed, and more than half of the soybeans in theUnited States were GMO's, soybean meal may also become unacceptable. Alfalfa is an excellent source of nitrogen and nutrients, but the price is not supportive of large scale use as a fertilizer.

One of the best presentations of the organic fertilizer options listed above is in the book Fertile Soil, A growers guide to Organic and Inorganic Fertilizers, by Robert Parnes (1990). Other useful books include "Solar Gardening" by Leandre and Gretchen Poisson (1994, Chelsea Green Publishing Company, White River Junction, Vermont) and "The New Organic Grower" by Elliot Coleman (1995, also by Chelsea Green Publishing).

PLANT NUTRIENTS AND THEIR FUNCTIONS

The plants require, the following essential nutrients for their normal development:

Carbon	Nitrogen	Calcium
Hydrogen	Phosphorous	Magnesium
Oxygen	Potassium	Sulphur
Iron	Zinc	Chlorine
Manganese	Boron	..
Copper	Molybdenum	..

Carbon is obtained from carbon dioxide of the air; Oxygen from air and water; Hydrogen from water; Nitrogen from air and soil or both, and all other nutrients from the soil. Soil is a the most important source of plant food.

Nitrogen, Phosphorous and Potassium are known as primary plant nutrients; Calcium, Magnesium and Sulphur are secondary nutrients; Iron, Manganes, Copper, Zinc, Boron, Molybdenum and Chlorine as trace elements or micronutrients. The primary nutrients and secondary nutrients elements are known as major elements.

This classification is based on their relative abundance, and not their relative importance. the micronutrients are required in small quantities, but they are as important as the major elements in plant nutrients.

Air is the primary source of Nitrogen for plant nutrient. Only leguminous crops can directly use this free Nitrogen with the help of symbiotic bacteria of the genus *Rhizobium*. Other plant derive from soil their Nitrogen in the form of Nitrogen and Ammonium. Nitrogen and Ammonium are produced in the soil by action of micro-organisms on the soil organic matter. Non symbiotic micro-organisms can fix free Nitrogen of the air and make it available to plant in Ammonium and Nitrates forms. *Nitrogen* encourages the vegetative development of plants by importing a healthy green colour to the leaves.

It also controls, to some extent the efficient utilization of phosphorous and Potassium. Its dependency retards growth and root development, turns the foilage yellowish or pale green, histens maturity, causes the shrivelling of grains and lowers crop yield. The older leaves are affected first. An excess of Nitrogen produces leathery(sometimes crinkled), dark-green leaves and succulent growth. It also delays the maturation of plants, impairs the quality of crops like barley, potato, tobacco, sugarcane, and fruits; increases susceptibility to diseases and causes 'lodging' of cereal crops by inducing an undue lengthoning of the stem internodes.

Phosphorous influences the vigour of plants and improves the quality of crops. It encourages the formation of new cells, promotes root growth(particularly the development of fibrous roots), and hastens leaf development through emergence of ears, the formation of grains, and the maturation of crops. It also increases resistence to diseases and strengthens the stems of cereal plants, thus reducing their tendency to lodge. It offsets the harmful effects of excess nitrogen in the plant.

When applied to leguminous crops, it hastens and encourages the development of nitrogen-fixing nodule bacteria. If phosphorous is deficient in the soil, plants fail to make a quick start, do not develop a satisfactory root-system, remain stunted and sometimes develop a tendency to show a reddish or purplish discolouration of the stem and foilage owing to an abnormal increase in the sugar content and the formation of anthoscyanin.

However, the deficiency of this element is not easily recognised as that of nitrogen. It has also been observed that cattle feeding on the produce of deficient soils become dwarfed, develop stiff joints and lose the velvetty feel of the skin. Such animals show an abnormal craving for eating bones and even soil itself.

Potassium enhances the ability of plants to resist diseases, insect attacks, and cold and other adverse conditions. It plays an essential part in the formation of starch and in the production and translocation of sugars, and is thus of special value to carbohydrate-rich crops, *e.g.* sugarcane, potato and sugar-beet.

The increased production of starch and sugar in legumes fertilized with potash benefits the symbiotic bacteria and thus enhances the fixation of nitrogen. It also improves the quality of tobacco, citrus etc. With an adequate supply of potash, cereals produce plump grains and strong straws. But an excess of element tends to delay maturity, though, not to be the same extent as nitrogen.

Plants can make up and store potassium in much larger for correcting zinc deficiency. The symptoms of zinc deficiency appear generally in younger leaves, starting with intervienal chlorosis leading to a reduction in shoot growth and the shortening of internodes. Mottle leaf, little leaf, etc. in the case of trees are symptoms of zinc difficiency. The buds of several defficient maize plants become white; in citrous interveinal chlorosis and mottled leaf occur. In calcareous soils and in soils with very high phosphorus content, zinc defficiency is commonly expected to occur.

The principal function of zinc in plants is as a metal activator of enzymes. In highly weathered coarse textured soils zinc deficiency appears under an intensive cropping programme. The availability of zinc is least between pH 5.5 and 7, but its availability increases at a lower pH. At a higher pH above 7, zinc availability becomes a complex problem, as the positively charged zinc ion gets converted into a negatively charged zincate complex whose availability tends to be reduced in alkaline soils.

When the calcium ion is predominent, the highly insoluble calcium zincate is formed and zinc availability gets seriously limited. The application of soluble zinc salts or zinc chelates to the soil is generally recommended to correct its defficiency. Foliar sprays are advocated, especially for orchard trees for amending zinc defficiency. About 5 to 50 kg of zinc sulphate per hectare is used for such purposes.

The symptoms of *boron* defficiency vary with the kind and age of the plant, the conditions of growth and the severity of the deffiency. Each crop produces its characteristic growth abnormalities associated with boron defficiency, such as yellows and rosetting in lucerne, snakehead in wallnuts, die-back and corking of fruits in apple, corking and pitting of fruits in tomatoes, hollow stem and the bronzing of curd in cauliflower, the brown-heart diseases in table-beets, turnips, etc.

Molybdenum deficiency produces whip-tail in cauliflower, broccoli and other *Brassica* crops. The deficiency of this element reduces the activity of the symbiotic and non-symbiotic nitrogen-fixing micro-organisms. It was in 1954 that *chlorine* was proved to be an essential micronutrient. Its defficiency under field conditions has not been reported so far. In water-culture solutions, the leaves of chlorosis, necrosis and an unusual bronze discolouration on tomatoes.

Sodium is not an essential element for plant growth. But some crops, such as beet, celery, cabbage, kale, knol-khol, radish, rape and turnip, benefit greatly by application of soluble sodium salts, specially if the soil is deficient in potassium. Sodium is also of direct benefit to plants indigneous to the sea-shore or to irrigated arid regions.

Salts of this element are said to release more of potassium from the exchange complex and to help to maintain phosphorus in a more available form. They also serve as a partial substitute for potassium in the case of potatoes and cotton.

APPLICATION OF FERTILIZERS AND MANURES

Bulky organic manures should be applied well ahead of sowing, so that the preliminary decomposition takes place before the seeds germinate. Failing presowing application, they may be applied any time after the seedlings have established themselves.

They are best applied in the powedered form. A sufficient supply of moisture in the soil is essential for their rapid decomposition. In the case of inorganic fertilizers, potassic and phosphatic fertilizers are best applied just before sowing or transplanting. Nitrogenous fertilizers may be applied either at planting and partly later. Split application is particularly desirable for nitrogen when applied to irrigated crops or to crops in heavy-rainfall areas.

Fertilizers applied before sowing should be broadcast uniformly and harrowed in. In the case of fertilizers containing soluble phosphate, the desirability of applying them in 2.5 to 5 cm wide bands on each side of the row of seeds at a depth of 10 to 15 cm with a drill has already been pointed out.

This operation reduces the fixation of soluble phosphate in the soil. It is also a good practice to mix superphosphate with farmyard manure at 18 to 22 kg to a tonne before applying the organic manure, particularly of dairy farms land. Sulphate of ammonia used as a top-dressing should not be applied when plant leaves are wet.

In the case of irrigated crops, the application of fertilizer should invariably be followed by a watering. In the case of fruit-trees, the fertilizer should be applied to the soil under the crown, a few metres away from the trunk. The area of application should be progressively extended as the trees grow bigger.

In advanced countries, fertilizers are usually applied with the help of machinery of diverse kinds and sometimes with an aeroplane or a helicopter. Combined-planters and fertilizer-distributors are employed when row crops are fertilized at sowing time.

DIAGNOSING THE FERTILIZER NEEDS OF SOILS

There are four methods of determining the fertilizer requirements of soil:

1. Field experiments,
2. Pot tests,
3. Biological tests,
4. Chemical test.

Field experiments contribute the more relaible method, but being time-consuming and expensive, they are conducted mainly by the research farms and research organisations. Farmers wishing to use field experiments as a Valid means of determining the fertility status of their soils should seek the advice of the state agronomist. Improperly conducted field experiments not only mean economic fertilizer practices.

Pot experiments permit test witha alarge number of manurial treatments within a limited space and in a relatively short time. However, as the conditions of such tests are different from those in the field, the results are not always directly applicable to large-scale farming.

Biological tests involve the growth of seedlings or of lower forms of plants, such as fungi and bacteria, under specified conditions and the study of their relative growth or the content of needed nutrients. A peridic testing of plant tissues for nitrates and other nutrients indicates the changing needs of crops for different food elements. But these are slow and costly processes and hence not always practicable.

The chemical analysis of soils or of plants growing on them constituents the modern method of determining the fertility status of a soil. Such analysis give information on the relative abundance or scarcity of the different nutrients required by crops from the soil, but they give no indication regarding the exact quantity, of fertilizer that may be applied to make good the deficiency. The dependability of this, method can, however, be increased a great deal by co-ordinating its results with those obtained from field experiments. Facilities for rapid soil-testing have been made available in almost all states and can be availed.

At the same time, a very large number of fertilizer experiments with different crops on a variety of soils are conducted annually in all parts of the country under a comprehensive scheme. The results of these field experiments, when calibrated against those of rapid soil tests, will make the latter purely dependable. This will then be a valuable aid in the hands of extension workers in furnishing advice to farmers regarding fertilizer practices. It is also sometimes possible to obtain a clue to the nutrient deficiences of soil with the help of deficiency symptoms in plants, as described already. However, a correct deagnosis of deficiency symptoms needs extensive experience. Furthermore, such symptoms in plants appear long after the actual occurance of nutrient deficiency in the soil. Therefore, such soil deficiencies must be diagnosed and remedied much earlier by adopting other means.

How much Fertilizer to Apply

In a vast country, such as India, possessing a wide variety of climatic and soil conditions, no definite quantity of any fertilizer can be prescribed as an optimum dose even for one and the same crop for all regions. each state in the country has conducted fertilizer investigations for the last 50 years and accumulated information on the manurial requirements of principal crops under local conditions. Therefore, for specific fertilizer practices in relation to particular crops and soils, the state agricultural department should always be consulted.

For determining the quantities of fertilizers from the recommended rates of application N,P or K, or *vice versa*, the following conversion factors may be used:

Table. Conversion Factors Determining Quantities of Fertilizers

Qunatity	Multiplied	Gives Corresponding by Quantity of
Nitrogen	4.854	Ammonium sulphate
Nitrogen	2.222	Urea
Nitrogen	3.846	Ammonium sulphate nitrate
Nitrogen	4.000	Ammonium chloride
Nitrogen	3.030	Ammonium nitrate
Phosphoric acid(P_2O_5)	6.250	Superphosphate, single
Phosphoric acid(P_2O_5)	12.222	Superphosphate, double
Phosphoric acid(P_2O_5)	2.857	Dicalcium phosphate
Phosphoric acid(P_2O_5)	5.000	Bone meal, raw
Potash (K_2O)	1.666	Muriate of potash
Potash (K_2O)	2.000	Sulphate of potash
Ammonium sulphate	.206	Nitrogen
Sodium nitrate	0.155	Nitrogen
Urea	0.450	Nitrogen
Ammonium sulphate nitrate	0.260	Nitrogen
Ammonium chloride	0.250	Nitrogen
Ammonium nitrate	0.330	Nitrogen
Superphosphate, double	0.450	Phosphoric acid(P_2O_5)
Dicalcium phosphate	0.350	Phosphoric acid(P_2O_5)
Bone meal, raw	0.200	Phosphoric acid(P_2O_5)
uriate of potash	0.600	Potash (K_2O)
sulphate of potash	0.500	Potash (K_2O)

Considerations Governing the use of Fertilizers

It may be stressed that in order to secure the maximum response to fertilizers, the crop should be irrigated immediately, following their application and at suitable intervals thereafter. The response of a crop under arid and semi-arid conditions is usually uncertain and relatively small. It is, therefore, advantageous to restrict the use of chemical fertilizers mainly ti irrigated lands and areas of assured rainfall.

It must also be borne in mind that the maximum profit from the use of fertilizers depends on many factors, such as the nature of soil, the kinds of crop grown, the climate (in relation to soil, the kinds of crops growth), the prices of fertilizers, the market price of the agricultural produce and so no. All these factors must be given due consideration to secure the most economic results.

Changes in one or more of them are bound to influence the economic results. changes in one or more of them are bound to influence the economics of fertilizer use. The use of fertilizers is also subject to the 'law of diminishing returns'. This means that the rate of increase in the crop yield decreases after a certain point is reached regarding the quantity of fertilizer used, and consequently the value of the additional yield finally becomes less than the cost of the fertilizer.

Usually, the smaller applications of fertilizers produce greater percentage increase of yield than larger applications. It should also be exphasized that an adequate supply of nutrients is only one of the factors that determine crop yield and that the application of fertilizers is not the sole means of making good the nutrient deficiencies in soils and plants. Equal attention must be paid to the other soil and crop-management practices to ensure good tilt, proper drainage, the required soil reaction, soil conversion, good land use, a suitable crop rotation, adequate organic matter in the soil and satisfactory soil micro-organism activity.

Each one of these plays a vital role in determining the eventual productioin. the neglecting of one or more or these factors leads to a reduction in yield and created the need for still heavier manuring. Finally, manuring should not only be balanced in itself, but should also be designed to supplement the good effects of proper land use and beneficial soil management.

Table. Conversion Factors

		Distance
1 mile	=	1760 yards = 5,280 feet
1 metre	=	1.0936 yards = 39.37 inches
1 yard	=	0.9144 metre
1 foot	=	0.3048 metre
1 inch	=	2.54 centimetres
	Area	
1 acre	=	4840 sq yards = 43,560 sq feet
1 *guntha*	=	1/40th acre = 1089 sq ft = 33' x 33'
1 acre	=	0.4047 hectare
1 hectare	=	2.471 acres
1 sq mile	=	640 acres
	Weight	
1 long ton	=	2,240 pounds
1 short ton	=	2,000 pounds
1 metric ton	=	1000 kilograms = 2204.6 pounds
1 long ton	=	28 maunds(approx.)
1 1 maund	=	80 pounds(approx.)
1 metric ton	=	1.1023 short tons
1 metric ton	=	0.9842 long ton
1 short ton	=	0.9072 metric ton
1 long ton	=	1.0161 metric ton
1 pound(avoir)	=	453.6 grammes = 0.4536 kg
1 hundred-weight (cwt)	=	112 pounds = 50.8 kg
1 kilogramme	=	2.2046 pounds
	Weight per area	
1 lb per acre	=	1.12 kilogrammes per hectare
1 cwt per acre	=	125.6 kilogrammes per hectare
	Capacity	
1 gallon(Imp.)	=	4.5461 litres

1 gallon (U.S.)	=	3.7853 litres
1 gallon (Imp.)	=	0.1604 cft
1 litre	=	0.2200 gallon (Br.)
1 litre	=	2642 gallon(U.S.)
1 bushel	=	56 lb (potaoes)
1 bushel	=	56 lb (barley)
1 bushel	=	60 lb wheat
1 bushel	=	42 lb (oats)
1 gallon (Imp.) water weighs 10 lb		
1 cft of water weighs 62.3 lb		

MAINTAINING SOIL FERTILITY

MANAGEMENT PRACTICES TO MAXIMIZE NUTRIENT CYCLING & NUTRIENT-USE EFFICIENCY

Nutrient management can be defined as "efficient use of all nutrient sources" and the primary challenges in sustaining soil fertility are to:

- Reduce nutrient losses
- Maintain or increase nutrient storage capacity
- Promote recycling of plant nutrients
- Apply additional nutrients in appropriate amounts

In addition, cultural practices that support the development of healthy, vigorous root systems result in efficient uptake and use of available nutrients. Many cultural practices help accomplish these goals, including establishing diverse crop rotations, reducing tillage, managing and maintaining crop residue, growing cover crops, handling manure as a valuable nutrient source, composting and using all available wastes or byproducts, liming to maintain soil pH, applying supplemental fertilizers, and routine soil testing. These beneficial management practices have multiple effects on nutrient cycling and soil fertility, which make it important to integrate their use and examine their effects on the complete soil-crop system, rather than just a single component of that system. There are many good ways to farm, so different solutions or combinations of practices are appropriate for different systems to reach similar goals.

PHOSPHORUS FLOWS AND THE MINNESOTA RIVER

Phosphorus enrichment of surface waters is a major issue in some parts of Minnesota. In fresh water systems, P is usually the limiting nutrient for growth of algae and aquatic plants, so their growth is stimulated when P in runoff or eroded soil enters lakes and rivers. Algal blooms lead to accelerated eutrophication of surface waters and degradation of water quality. In extreme cases, depleted levels of dissolved oxygen in the water cause death of fish and other aquatic life. Water quality concerns led to a recent Minnesota law restricting P fertilizer application on home lawns and other turfgrass areas.

In the agricultural landscape, similar concerns are expressed about the role of agriculture in P enrichment of the Minnesota River. P loading into the Minnesota River comes from a variety of sources, including stream bank erosion, water treatment plants, and industrial activity. The extent of the contribution from agriculture is difficult to measure, but P in runoff from farm fields and P attached to eroded soil are certainly potential pathways of P delivery to the river and its tributaries.

Phosphorus flows into Minnesota are not as dramatic as those described for the poultry production areas of the Delmarva Peninsula, but there are similarities and some common pathways. Phosphate rock mined in Florida or other distant locations is processed into phosphate fertilizers that are transported to Minnesota and applied to crop fields. Some of the grain harvested from these fields becomes part of animal feeds that are shipped to places like the Delmarva Peninsula, and additional harvested products are transported for other uses, so some of the imported P flows back out of the state in exported agricultural products. However, some of the imported P accumulates in various forms and locations, and is a potential source of nutrient loading and impaired water quality in the Minnesota River if not properly managed.

Accumulation of P in manure and increasing levels of P in the soil are two ways the flow of P into the state can build up and threaten the Minnesota River or other surface water bodies if not managed efficiently. Concentrated livestock production is not as widespread as it is in the Delmarva Peninsula, but in localized areas the amount of P in manure exceeds the amount of P required to meet crop needs on surrounding cropland. Manure is a valuable resource, but fields with a long history of heavy manure application can exceed the capacity of the soil to efficiently recycle the amount of P in continual manure additions. Buildup of soil P can also occur when P fertilizer is applied at rates exceeding crop P requirements.

Efficient use of fertilizer and manure P requires sound nutrient management planning to reduce the potential for environmental problems. This includes soil testing to determine the need for P, manure analysis, proper storage and handling of manure, and fertilizer and manure application methods that reduce the potential for movement of P from farm fields. In addition, soil management practices that limit surface runoff and reduce soil erosion help protect water quality, as well as sustaining long-term soil productivity.

CROP ROTATIONS

The term 'rotation effect' was coined to describe the observation that yields for a crop grown in rotation with other crops are usually 5 to 15% greater than for continuous monoculture of that same crop. The reason for increased yields is not always clear, and in most cases it is probably not due to a single cause, but growing a variety of crops in sequence has many positive effects on soil fertility. In a diverse rotation, deep-rooted crops alternate with shallower, fibrous-rooted species to bring up nutrients from deeper in the soil. This captures nutrients that might otherwise be lost from the system. Differences in plant rooting patterns, including root density and root branching at different soil depths, also results in more efficient extraction of nutrients from all soil layers when a series of different crops is grown.

Including sod-forming crops in rotation with row crops decreases soil and nutrient losses from runoff and erosion, and increases soil organic matter. Growing legumes to fix atmospheric N reduces the need for purchased fertilizer and increases the supply of N stored in organic matter for future crops. Biologically fixed N is used most efficiently in rotations where legumes are followed by crops with high N requirements. Rotating crops also increases soil biodiversity and nutrient cycling capacity by supplying different residue types and food sources, reduces the buildup and carryover of soil-borne disease organisms and insect pests (breaks disease and pest cycles), and can help create favorable growing conditions for healthy, well-developed crop root systems.

SOIL AND WATER CONSERVATION PRACTICES

Soil erosion removes topsoil, which is the richest layer of soil in both organic matter and nutrient value. Implementing soil and water conservation measures that restrict runoff and erosion minimizes nutrient losses and sustains soil productivity. Tillage practices and crop residue cover, along with soil topography, structure, and drainage, are major factors in soil erosion. Surface residue limits erosion by reducing detachment of soil particles by wind or raindrop impact and restricting water movement across the soil. Tillage practices manage the amount of crop residue left on the soil surface. Reduced tillage or no-till maximizes residue coverage. Water moves rapidly and is more erosive on steep slopes, so reducing tillage, maintaining surface residue, growing sod crops, and planting on the contour or in contour strips are recommended conservation practices. Using diverse rotations and growing cover crops also can reduce erosion.

Soils with stable aggregates are less erodible than those with poor structure, and organic matter (including the activity of living soil organisms and fine roots) helps bind soil particles together into aggregates. Tillage breaks down soil aggregates and also increases soil aeration, which accelerates organic matter decomposition. Well-drained soils with rapid water infiltration are less subject to erosion, because water moves rapidly into and through them and does not build up to the point where it moves across the surface. Drainage improvements on poorly drained soils reduce runoff, erosion, and soil compaction. Improving drainage also decreases N losses from denitrification, which can be substantial on waterlogged soils, by increasing aeration. Improving aeration in the plant-root zone also promotes healthy root growth. A negative consequence of improved drainage is loss of nitrate-N and other nutrients through tile outlets to surface waters. Especially important are flushes of residual N after late winter/early spring rains.

COVER CROPS AND GREEN MANURES

Growing cover crops and green manure crops can be viewed as a type of crop rotation, where adding a non-revenue generating crop between annual cash crops extends the growing season. Many of the benefits, therefore, are the same as those achieved with crop rotation.

The terms cover crop and green manure are frequently used synonymously. They perform many similar functions and many of the same plant species are used as both cover crops and green manure crops. The main difference between the two is that the primary purpose of growing a cover crop is to protect the soil surface from raindrop impact, runoff, and erosion and the primary purpose of a green manure is as a soil-building crop to produce organic material for incorporation into the soil. Winter grains like cereal rye planted after potatoes are cover crops that are designed to hold soil in place until the next main crop is planted in the spring, but they also add organic matter to the soil when they are turned under. Rapidly growing summer annuals like buckwheat and sorghum-sudangrass are planted between short-season vegetable crops as green manures to add organic matter to the soil, but they also protect the soil from erosion between spring and fall vegetables.

Growing legume cover crops adds biologically fixed N. The additional plant diversity with cover crops stimulates a greater variety of soil microorganisms, enhances carbon and nutrient cycling, and promotes root health. The soil surface is covered for a longer period of time during the year, so nutrient losses from runoff and erosion are reduced. This longer period of plant growth substantially increases the amount of plant biomass produced, which in turn increases organic matter additions to the soil. The extended growth period obtained with cover crops also extends the duration of root activity and the ability of root-exuded compounds to release insoluble soil nutrients.

A winter cover crop that makes good fall growth traps excess soluble nutrients not used by the previous crop, prevents them from leaching, and stores them for release during the next growing season. Complementary cover crop mixtures produce root exudates with varying composition and effects, and have different zones of nutrient uptake, because they differ in amount, depth, and patterns of root branching. Deep-rooting cover crops, like sorghum-sudangrass hybrids and sweet clover, can break up some types of compacted soil layers and improve rooting depth for the next crop. Cereal rye, sorghum-sudangrass, and brassicas (mustards), such as oilseed radish and forage turnip, all suppress some nematode species and may be useful cover crops in fields with moderate infestation levels. Cover crops also can suppress weeds, which otherwise would compete with crops for nutrients.

Cover crop benefits are probably greatest as soil-building crops preceding high-value perennial fruits and in rotations with low-residue, short-season crops such as annual vegetables. It is often easier find places to grow cover crops in vegetable rotations than in agronomic rotations, and there may be opportunities to grow both summer and fall cover crops in vegetable systems. Many vegetables have relatively shallow, sparse root systems, but are well fertilized because of their value. Both summer and fall cover crops absorb residual nutrients, in addition to increasing the time and amount of surface cover.

Disadvantages of growing cover crops are:

- Large amounts of residue can make planting difficult and reduce crop stands
- In wet springs, planting may be delayed if wet soil conditions delay killing the cover crop
- Soil warms more slowly in the spring under cover crops than for tilled soil and lower soil temperatures can slow seed germination, reduce early-season growth, delay maturity, and reduce crop yields
- Spring cover crop growth uses water, which can adversely affect the following cash crop in a dry year (in wet years, cover crop water use may be beneficial on poorly drained soils)
- Some cover crops attract and/or harbor pests that can damage succeeding crops. There are expenses and management time required to grow cover crops

Cover crops have many benefits, but when you grow them you need to commit time to their selection and management to fully realize their benefits and avoid potential problems. Select cover crops with characteristics that will meet your objectives and fit your rotations, and then manage them with the same attention and skill you give any other crop.

MANURE MANAGEMENT

Returning manure to crop fields recycles a large portion of the plant nutrients removed in harvested crops. On farms where livestock are fed large amounts of off-farm purchased feeds, manure applied to crop fields is a substantial source of nutrient inputs to the whole farming system. However, just as nutrients can be lost from the soil, nutrient losses from manure during storage, handling, and application are both economically wasteful and a potential environmental problem. Soluble nutrients readily leach from manure, especially when it is unprotected from rainfall during storage. N is readily lost through volatilization of ammonia, both during storage and when manure is not incorporated soon after field application. Nutrient losses from manure also occur when it is applied at rates exceeding crop nutrient requirements.

Nutrient management is the efficient use of all nutrient sources. A Nutrient Management Plan that takes all nutrient sources into account is not just environmentally sound, it is good business.

Analyze manure for its nutrient content and adjust application rates according to crop needs, soil tests, and frequency of manure applications. Avoid applying manure at rates that exceed crop requirements for any nutrient, but especially for N and P on fields that receive manure on a regular basis. This often means that rates should be based on P requirements rather than N requirements. Following heavy manure applications with crops that have high nutrient requirements (especially for N and P) reduces losses and increases nutrient-use efficiency. In addition to nutrient value, manure adds organic matter to the soil, which can improve soil structure and increase CEC. Refer to Using Manure and Compost as Nutrient Sources for Vegetable Crops for further information on nutrient content, nutrient availability, and calculation of application rates for efficient use of manure as a source of plant nutrients for vegetable crop production.

COMPOST AND OTHER SOIL AMENDMENTS

In addition to manure, organic amendments such as biosolids, food processing wastes, animal byproducts, yard wastes, seaweed, and many types of composted materials are nutrient sources for farm fields. Biosolids contain most of the essential plant nutrients, and are much "cleaner" than they were twenty years ago, but regulations for farm application must be followed to prevent the possibility of excessive trace metal accumulation. Biosolids are also not an acceptable nutrient source for certified organic production.

Composting is a decomposition process similar to the natural organic matter breakdown that occurs in soil. Proper composting conserves volatile and soluble N, and other mobile nutrients in waste products, by incorporating them into organic forms where they are more stable and less readily lost. Composting reduces the bulk of organic wastes and makes transportation and field application of many waste products more feasible. On-farm composting of manure and other farm wastes also facilitates their handling. Most organic materials can be composted, nearly all organic materials contain plant-nutrient elements, and recycling all suitable wastes or byproducts through soil-crop systems by either composting or direct field application should be encouraged. These practices build up soil organic matter and provide a long-term, slow-release nutrient source. Some composts also have disease-suppressive properties that improve root growth and health.

Inorganic byproducts also can be recycled through the soil and supply plant nutrients. Available materials vary by region, but wood ash, rock dust from quarries, gypsum from scrubbers in power plants burning high-sulfur coal, and waste lime from water treatment plants are among the waste products that are beneficially re-used. When considering the agricultural use of any byproduct, a thorough chemical analysis and review of possible regulations should be done to avoid soil contamination problems. Even seemingly benign byproducts should be analyzed and field-tested on a trial basis before using them on a large acreage.

Table . Percentage of the Total Soil Volume Occupied by Plant Roots (in the surface 8-inches of soil)

Crop	Root volume (%)
Kentucky Bluegrass	2.8
Winter Rye	0.9
Oat	0.6
Soybean	0.4-0.9
Corn	0.4

Note:

Adapted from S. Barber, *Soil Nutrient Bioavailability*, 1984

Healthy, vigorous root systems

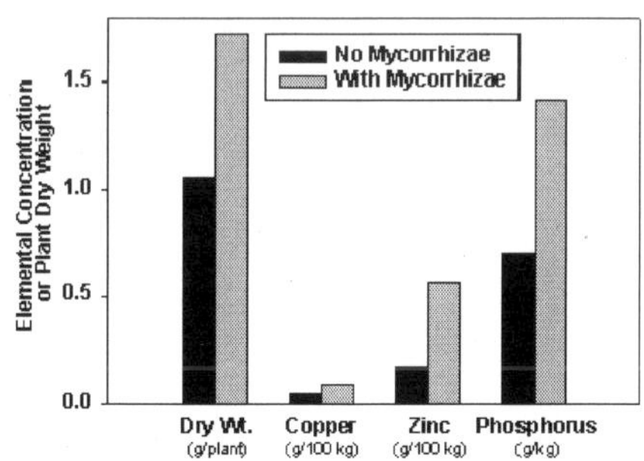

Adapted from S. Barber, *Soil Nutrient Bioavailability*, 1984 (from data of Lambert et al., 1979, Soil Sci. Soc. Am. J. 43:976-980).

Fig. Effect of mycorrhizae on growth and Cu, Zn, and P content of 40-day-old soybean plants

Vigorous root systems tap nutrient supplies from a larger volume of soil, so management practices that stimulate healthy root growth can also increase nutrient uptake. Uptake efficiency by extensive, well-distributed root systems results from increases in the amount of root surface area in contact with the soil. The extent of root-soil contact is limited by the fact that roots occupy only about 1 to 3% or less of total soil volume, even for fibrous-rooted plants in the surface layer of soil where root density is greatest. For immobile nutrients like P, root growth to the nutrient is especially important for uptake, because in most soils P moves only about 1/10 of an inch over the entire growing season.

Root-soil contact is determined by root length (both vertical & horizontal), root branching, and root hairs. Root hairs are located just behind the root tip and have a relatively short life span of a few days to a few weeks. Actively growing feeder roots are necessary to continually renew these important locations for nutrient uptake. Symbiotic associations between soil fungi and plant roots also increase nutrient absorbing capacity. These fungi, called mycorrhizae ("fungus roots"), function as an extension of plant root systems. Mycorrhizae obtain food from plant roots and in return increase the nutrient absorbing surface for the plant through their extensive network of fungal strands (hyphae). Mycorrhizae are particularly important for P uptake in low P soils. They can increase Zn and Cu uptake and also provide some protection against root disease.

Root activity also has direct effects on nutrient availability in the soil. Insoluble nutrients are released and maintained in solution by the action of organic acids, chelates, and other compounds produced by roots. Nutrients are also released because the soil immediately adjacent to roots, the rhizosphere, often has a lower pH than the bulk soil around it as a consequence of nutrient uptake. The rhizosphere stimulates microbial activity and microbes also release compounds like organic acids, enzymes, and chelates that solubilize nutrients.

A number of soil factors and management practices affect root growth, distribution, and health. Compacted soil layers restrict root penetration, low pH in the subsoil can restrict rooting depth, water saturation and poor aeration inhibit root growth, and roots will not grow into dry zones in the soil. Alleviating these conditions through some of the management practices described in this bulletin can increase nutrient uptake. Cultural practices that promote soil biodiversity help maintain healthy root systems, because an active and diverse microbial population competes with root pathogens and can reduce root disease.

APPLICATION OF ORGANIC MATERIALS ON SOILS USED FOR CROP PRODUCTION

Land application of organic waste material is a desirable disposal alternative. Not only are costs usually lower relative to other disposal methods but the waste material is beneficial for the soil and crop production. Organic materials contain nutrients needed for crop growth and also to improve soil tilth, increase water holding capacity, lessen wind and water erosion, improve aeration, and promote soil biological activity. Application of organic waste materials to land must, however, take into account both crop needs and the potential for environmental degradation. Whether land disposal of organic waste is viewed as a nutrient resource or as a waste disposal option, certain principles needed to be followed for optimum nutrient utilization and minimum environmental impact.

FACTORS WHICH INFLUENCE APPLICATION RATE

Nutrients in Organic Matter - Crop Growth Considerations

Most organic materials contain significant quantities of plant nutrients, especially phosphorus (P), nitrogen (N) and potassium (K), and should be managed as a mixed fertilizer to satisfy crop nutrient needs. Nutrients should not be applied in quantities that exceed the amount needed for optimum plant nutrition.

Because many Ohio soils have relatively high P fertility levels, efficient management of organic matter P is of primary concern. P readily accumulates in soils if applied in quantities greater than those removed by crops. Crops grown in soils with very high phosphorus levels may actually produce lower yields due to nutrient imbalances.

Phosphorus applied to fields can move into bodies of water during erosion and runoff events, and is largely responsible for the accelerated eutrophication of many bodies of water in Ohio. Concentrations of algal-available phosphorus (the phosphorus responsible for eutrophication) increase as soil test phosphorus levels increase, meaning that as soil concentrations of phosphorus rise, the potential increases for soils to degrade the environment through runoff and erosion. Slowing the eutrophication of Ohio's surface waters involves reducing the amount of phosphorus delivered to these waters.

Agronomic crops grown in Ohio rarely respond to applications of additional phosphorus when soil test levels exceed 30 ppm (60 lb/acre) of phosphorus. There is no agronomic justification for raising soil test phosphorus levels above those that provide adequate nutrition to the crop. Increasing soil test phosphorus levels at the soil surface increases the pollutant potential of the land.

High levels of organic materials applied to soil may increase salts in soil, impair seed germination and reduce plant stands. Most organic materials also contain much more potassium than magnesium or calcium. Long term application of this material to soils may raise the ratio of potassium to magnesium and calcium sufficiently to retard crop growth. To adjust the ratio, additional magnesium and/or calcium may have to be added as dolomitic or calcitic limestone. When organic materials are regularly applied to cropland the soil should be tested regularly to monitor nutrient levels. If a mineral imbalance is suspected, plant tissue analysis should be made to determine the extent of the problem.

Applications of organic materials should not provide more available nitrogen (N) than is needed by the succeeding crop. The determination of total available nitrogen should include credits for any contributions of the present or preceding crop, any nitrogen fertilizer added, and available nitrogen provided by previous waste applications.

Method of waste application to the soil can affect the amount of nutrients available for crop uptake. Availability of N in organic waste depends both on the method of application and days to incorporation into the soil. Organic materials should be incorporated into the soil as soon as possible since most N losses occur within 24 hours of application. Injecting, chiseling, or knifing liquid materials into the soil minimizes odors and nutrient losses. Phosphorus and potassium contained in organic waste is generally readily available in the soil unless removed by surface runoff or soil erosion.

When planning for land application of organic wastes, sufficient land area to accommodate the nutrient supply must be considered. A conservative approach in determining the amount of land required is to base application rate on the nutrient removal of the harvested crop. This approach will prevent nutrient buildup in the soil beyond suggested agronomic production levels. For information on determining required land area for waste application based on crop nutrient removal, refer to "Land Required for Manure Application Based on Crop Removal" in OSU Extension Bulletin 604 (*Ohio Livestock Manure and Wastewater Management Guide*).

Organic material should be tested prior to application since nutrient composition of organic waste is affected by collection, storage and the waste handling system. Soils receiving organic materials should also be tested for available nutrients before application. Nutrients supplied from commercial fertilizer and/or manure sources must also be considered in calculating the appropriate application rate. The maximum organic material application rates for different soil test levels and site conditions. A computer program available from Ohio State University Extension can also assist in selecting appropriate application rates. ECP 102, *Crop Nutrient Management*, is available from your local county Extension office.

NUTRIENTS IN ORGANIC MATTER - ENVIRONMENTAL CONSIDERATIONS

Site characteristics, application methods and management, cropping system, and erosion/runoff abatement practices all affect the quantity of waste that can be safely applied to a given site. Since most of the nutrients contained in waste materials are water soluble, the potential for these nutrients to cause surface water pollution is directly related to the level of water runoff from the application site. Waste organic matter particles are also subject to movement in surface water runoff and can contribute directly to pollution of surface water supplies.

Water and nutrient runoff potential are affected by numerous factors, some of which are fixed by the nature and location of the field and others that can be altered through management. Runoff potential must be determined on a site-by-site basis.

FACTORS AFFECTING RUNOFF POTENTIAL

Proximity of surface water to application site. Streams or bodies of surface water located near waste application sites are more likely to receive waste derived nutrients. The extent and management of buffering areas such as other fields, pasture, wooded area, etc. between the source of nutrients and the water supply can greatly attenuate the movement of nutrients to the water supply.

Slope steepness and complexity. Runoff is more likely from fields sloping steeply and evenly toward a water source than fields with a gentle or no slope. Fields with areas of depression between the site of application and the water source have a lower potential for nutrient runoff.

Soil and weather conditions. Organic materials should not be applied to wet, sloping, or frozen soils if normally anticipated rainfall would cause overland water flow from the point of applications. Liquid organic sources should not be applied at rates that exceed the volume needed to bring the soil to field moisture capacity. Check soil moisture before applying liquid wastes and adjust application rates to avoid runoff. Estimate soil moisture based on soil feel and appearance. Refer to EX 704, *Land Application of Manure and Waste Water - Part I.* Available moisture capacity for different soils can be found in the Ohio Irrigation Guide available from the state Soil Conservation Service or the Department of Agricultural Engineering, The Ohio State University.

Soil type and surface condition. Soils with low infiltration rates and/or soils with limited water-holding capacity are more likely to promote water runoff than soil types that absorb and retain large quantities of water. A rough or covered soil surface also reduces runoff compared with soil surfaces that are smooth or have very little residue cover.

Subsurface drainage. Subsurface drainage systems can reduce the water runoff and allow the nutrients contained in the waste to be absorbed into the soil profile.

Waste application method. Surface spreading and subsurface injection are the most common land-application methods. Injection or incorporation by primary tillage promptly following application will reduce the potential for direct runoff as well as reduce odors and volatile N loss.

Application equipment calibration. It is important that equipment for applying waste be properly calibrated and be able to uniformly apply waste material.

LAND DISPOSAL OF ORGANIC MATTER WASTE

When organic materials are applied in excess of the amount that can be effectively utilized based on crop nutrient needs, special guidelines should be considered. For instance, in many situations, relatively deep incorporation of the organic matter and control of erosion and surface water runoff at the application site can greatly reduce environmental impacts.

If organic materials must be applied to soils with Bray P_1 levels greater than 30 ppm (60 lb/acre) of phosphorus in the top 8 inches, the following recommendations should be considered:

No additional phosphorus fertilizer should be used.

Crops should be monitored for nutrient deficiencies using plant tissue analysis. Increasing soil test phosphorus and potassium levels above recommended levels increases the probability of yield reducing nutrient imbalances.

Organic materials should be applied in quantities such that the long-term phosphorus level at the soil surface does not increase appreciably. This can be accomplished by:

a. Apply no more nitrogen or phosphorus (whichever is lower) than will be removed by the next crop (one season) if surface-applied or incorporated to a shallow depth (within the tillage depth).
b. Incorporate waste below the depth of tillage, generally deeper than 8 inches, using rates great enough to satisfy nitrogen requirements for a succeeding grass crop. Producers should be aware that this system may promote development of a soil zone with abnormally high nutrient concentrations. If this zone is brought to the surface by subsequent tillage, potential for P runoff is increased. Fields under such management should be sampled deep enough to include the zone of incorporation. On soils with high nutrient levels, application of organic materials should be skipped for one or more seasons to allow depletion of accumulated nutrients.

Soil and water conservation practices should be implemented to control erosion and minimize runoff.

Application of organic material that contains P is not recommended for crop production where the Bray P_1 level in the top 8 inches of soil exceeds 300 pounds of phosphorus per acre.

If deep application and incorporation of organic material are required to insure low runoff and more than 250 pounds of fertilizer P_2O_5 per acre are required for crop production, soil at the depth of application should not be brought to the surface for three years unless low runoff conditions can be maintained. Moldboard plow depth should be set to two-thirds the depth of incorporation during this three year period, and non-inversion tillage tools such as chisel plows should be set no deeper than the depth of application. Deep incorporation of less than 250 pounds of P_2O_5 per acre may be conventionally tilled the following year.

DETERMINATION OF NUTRIENT AVAILABILITY OF ORGANIC FERTILIZERS FOR POTENTIAL USE ON GREENHOUSE CROP

The greenhouse industry depends primarily on the use of synthetic (inorganic, organic) fertilizers for the supply of the major and minor plant nutrients for both vegetable and ornamental production. These fertilizers are often in a soluble, immediately plant available form, so that they can be provided with the irrigation water, i.e. fertigation, in a timely fashion and at the concentrations optimal for plant growth at each particular growth stage. There are both single nutrient, for example urea, as well as multi-nutrient fertilizers for this purpose (e.g. potassium nitrate).

Due to public pressure and the recent recommendation by the Ontario floricultural research and services subcommittee (OGFRSS), there is a need for the greenhouse industry to become more sustainable and one of the components of sustainability is plant nutrition. The focus of this study was to determine whether organic products, typically organic waste products from agriculture and other industries producing and processing food products, could be used to provide nutrients for plant production in a greenhouse setting. The main goal of the research was to use materials, which are in ample supply and not chemically modified in any way.

The study consisted of 2 parts: (i) to determine the nutrient availability of various organic fertilizers over time in a number of standard potting mix substrates; and (ii) to apply the obtained knowledge on the production of an 'organic' crop of potted plants. This report deals with the first objective only, for a small number of organic materials.

MATERIALS AND METHODS

The first part of the study determined the plant nutrient availability in standard substrates (sphagnum peatmoss; coco coir; granular rockwool; and soil) amended with a range of organic fertilizers. Although many organic fertilizers were used, we only report on the results of alfalfa (5-1-5), blood meal (12-0-0), corn meal (cornPlus; 8-1-6), and feather meal (12.5-0-0). The organic materials were obtained from Vineland Growers' Coop, Ltd., Jordan Station, ON. The peat moss was a commercial peat amended with some perlite, gypsum and limestone to create a pH of about 6.0 (Sunshine #2 basic mix, SunGro Horticulture, Bellevue, WA) ; the coco coir was a commercially available material (1 L compressed bricks, source unknown), which was soaked with DI water overnight for expansion; granular rockwool was a commercially available product (Plant Products Co., Bramalea, ON); while the top soil (sandy loam) was obtained from the Vineland Experimental Station.

All substrates were pre-wetted and the moisture contents were (g of water/ 100 g of dry substrate): peat:160; coco coir: 450; rockwool:100; and soil:10. After moistening, the substrates were mixed with the organic fertilizers at a rate of 0.5 kg of N/m^3, and placed in plastic bags and sent to a laboratory for analyses.

At the lab, two extraction techniques were used for comparison namely, the saturated medium extraction (SME), a technique commonly used in the greenhouse industry, and the field testing method. For the SME technique, DI water is added to the substrate until the substrate is saturated after which the solution is extracted and analyzed for nutrients (NO_3^-- N, ortho-P, K, Ca, Mg, S, Na, Cl, pH and electrical conductivity (EC)). The field testing method uses a number of different extraction solutions specific for each of the different nutrients in order to provide an index of what is available over the duration of a field crop (e.g. KCl for nitrogen; bicarbonate for phosphorous and ammonium acetate for K and Mg).The samples were sent to different labs: Agri-Food laboratories for the SME test and the soil testing lab at University of Guelph for the field soil test.

Analyses for the SME test were done on different days (day 1, 8, 22 and 43 after mixing), while the field test analyses were done 2 months after mixing and again 4 months later. The substrate/organic fertilizer samples were kept at room temperature and a new sample was taken from the substrate.

RESULTS

From a production point of view, the availability/ mineralization of the organic fertilizers into ammonium (NH_4^+) and nitrate (NO_3^-) are likely the most important parameters. The results of the field analyses (column 3-8) and SME (column 9-10).

Field analyses

The results of the substrate means showed that the ammonium levels decreased in all substrates (except rockwool) while that of NO_3^--N increased over time over a period of 16 weeks. These results can be attributed to mineralization (break-down of organics) and nitrification (conversion from ammonium to nitrate).

The rate of decrease in NH_4^+-N was not the same for all substrates. On a weight basis, the greatest mineralization took place in the peatmoss with relatively low levels of mineralization in rockwool and coco coir. It should be stressed that there was a strong interaction between the substrate and the organic fertilizers in terms of mineralization. The mean initial ammonium levels of the alfalfa, blood meal, corn plus and feather meal combined over all substrates were about 1430 mg of NH_4^+-N, and about 73 mg of NO_3^--N/ kg of substrate. The results after 16 weeks were about 966 mg of NH_4^+-N/ kg, and about 667 mg of NO_3^--N/ kg of substrate. This showed that the decrease in ammonium-N was in line with an increase in NO_3^--N. It should be noted that the results of the field analyses were all expressed in mg of N/ kg of dry weight of the substrate, while the supplementation was based on a volume basis. If one wanted to convert the results to mg of N/ L of substrate, then these numbers would have to be multiplied with the bulk density of the substrates, which are in g/ L (approximately): peat moss (100); coco coir (100); rockwool (50) and soil (1500). If we then look at mineralization on a volume basis, then the soil showed the highest mineralization followed by peatmoss, and negligible amounts for coco coir and rockwool. If we consider the increase in total nitrogen (columns 7+8), then soil increased by about 57 mg N/L, while peatmoss increased by 26 mg N/L, while coco coir and rockwool did not increase in total N. Alfalfa, feathermeal and 'cornPlus' all showed a similar increase in NO_3^--N levels (750 mg N/kg), if we compare the amount ofNO_3^--N at 6 months with levels at 2 months. Bloodmeal showed a considerably lower level of mineralization (415 mg N/kg of substrate over the same period of time.

Saturated Medium Extraction:

Unfortunately, the ammonium levels were not analyzed by the SME tests. The initial NO_3^-- N levels in all substrates without any organic fertilizer amendment (control) except for soil showed no presence of NO_3^-- N. Soil was the only substrate that showed high levels of mineralization. After 9 weeks, alfalfa, feather meal, cornPlus and blood meal showed increased NO_3^--N levels of about 400 ppm N compared to the day of mixing with soil (the latter is close to what was applied in total (500 mg of N/L). Peat moss also showed some mineralization but was less than that for soil. Rockwool and coco coir showed no significant amount of mineralization.

The results of some other nutrient elements such as pH, P and EC.

The initial pH of rockwool was the highest (8.0), while the pH's of peatmoss and coco coir were about 6.0. The soil-pH was at about 7.0. After the organic fertilizers were added to the substrates, the pH's decreased compared to that of the control. The greatest pH drop was found with rockwool (1.5 unit), but it increased again back to that of the control. It is unclear why the pH's in the field tests are showing different results from the SME method. The phosphorous levels in the rockwool increased but decreased over time in peatmoss or coco coir.

The EC levels (mS/cm) in the soil increased from 1.9 (t=0) to 5.7(t=9 weeks) for the substrates mixed with organic fertilizers. This increase in EC is partly due to the increase in inorganic N, but also due to the increased levels of calcium and sulphates.

The field soil testing method provided different results from the saturated medium extraction method for the rate of mineralization based on actual absolute values. The main reason is that field test reports are based on a per weight basis. So substrates with a high bulk density show low rates of mineralization. If data are expressed on a per volume basis, both the field test method and SME technique showed that the mineralization of nitrogen is the greatest in soils (about 80% of the applied organic nitrogen was converted to NO_3^--N), and that there was no mineralization for rockwool or coco coir. Mineralization in peatmoss, however, proved to be still substantial. It has become clear that the only non-processed medium (soil) showed the greatest mineralization and subsequent nitrification of the ammonium. Of the 4 organic fertilizers presented here, alfalfa showed the quickest release of nitrogen (within 10 days), but after 6 weeks, there were no differences between the 4 organic fertilizers.

PLANT NUTRITION

Plant nutrition is the study of the chemical elements and compounds that are necessary for plant growth, and also of their external supply and internal metabolism. In 1972, E. Epstein defined two criteria for an element to be essential for plant growth:

1. In its absence the plant is unable to complete a normal life cycle; or
2. That the element is part of some essential plant constituent or metabolite.

This is in accordance with Liebig's law of the minimum. There are 14 essential plant nutrients. Carbon and oxygen are absorbed from the air, while other nutrients including water are obtained from the soil. Plants must obtain the following mineral nutrients from the growing media:

- The primary macronutrients: nitrogen (N), phosphorus (P), potassium (K)

- The three secondary macronutrients: calcium (Ca), sulfur (S), magnesium (Mg)
- The micronutrients/trace minerals: boron (B), chlorine (Cl), manganese (Mn), iron (Fe), zinc (Zn), copper (Cu), molybdenum (Mo), nickel (Ni)

The macronutrients are consumed in larger quantities and are present in plant tissue in quantities from 0.2% to 4.0% (on a dry matter weight basis). Micro nutrients are present in plant tissue in quantities measured in parts per million, ranging from 5 to 200 ppm, or less than 0.02% dry weight.

Most soil conditions across the world can provide plants with adequate nutrition and do not require fertilizer for a complete life cycle. However, humans can artificially modify soil through the addition of fertilizer to promote vigorous growth and increase yield. The plants are able to obtain their required nutrients from the fertilizer added to the soil. A colloidal carbonaceous residue, known as humus, can serve as a nutrient reservoir. Even with adequate water and sunshine, nutrient deficiency can limit growth.

Nutrient uptake from the soil is achieved by cation exchange, where root hairs pump hydrogen ions (H^+) into the soil through proton pumps. These hydrogen ions displacecations attached to negatively charged soil particles so that the cations are available for uptake by the root.

Plant nutrition is a difficult subject to understand completely, partly because of the variation between different plants and even between different species or individuals of a givenclone. An element present at a low level may cause deficiency symptoms, while the same element at a higher level may cause toxicity. Further, deficiency of one element may present as symptoms of toxicity from another element. An abundance of one nutrient may cause a deficiency of another nutrient. For example, lower availability of a given nutrient such as $SO_2^{4"}$ can affect the uptake of another nutrient, such as NO_3^-. As another example, K^+ uptake can be influenced by the amount of NH_4^+ available.

The root, especially the root hair, is the most essential organ for the uptake of nutrients. The structure and architecture of the root can alter the rate of nutrient uptake. Nutrient ions are transported to the center of the root, the stele in order for the nutrients to reach the conducting tissues, xylem and phloem. The Casparian strip, a cell wall outside of the stele but within the root, prevents passive flow of water and nutrients, helping to regulate the uptake of nutrients and water. Xylem moves water and inorganic molecules within the plant and phloem accounts for organic molecule transportation. Water potential plays a key role in a plants nutrient uptake. If the water potential is more negative within the plant than the surrounding soils, the nutrients will move from the region of higher solute concentration—in the soil—to the area of lower solute concentration: in the plant.

There are three fundamental ways plants uptake nutrients through the root: 1.) simple diffusion, occurs when a nonpolar molecule, such as O_2, CO_2, and NH_3 follows a concentration gradient, moving passively through the cell lipid bilayer membrane without the use of transport proteins. 2.) facilitated diffusion, is the rapid movement of solutes or ions following a concentration gradient, facilitated by transport proteins. 3.) Active transport, is the uptake by cells of ions or molecules against a concentration gradient; this requires an energy source, usually ATP, to power molecular pumps that move the ions or molecules through the membrane.

- Nutrients are moved inside a plant to where they are most needed. For example, a plant will try to supply more nutrients to its younger leaves than to its older ones. When nutrients are mobile, symptoms of any deficiency become apparent first on the older leaves. However, not all nutrients are equally mobile. Nitrogen, phosphorus, and potassium are mobile nutrients, while the others have varying degrees of mobility. When a less mobile nutrient is deficient, the younger leaves suffer because the nutrient does not move up to them but stays in the older leaves. This phenomenon is helpful in determining which nutrients a plant may be lacking.

Many plants engage in symbiosis with microorganisms. Two important types of these relationship are 1.) with bacteria such as rhizobia, that carry out biological nitrogen fixation, in which atmospheric nitrogen (N_2) is converted into ammonium (NH_4); and 2.) with mycorrhizal fungi, which through their association with the plant roots help to create a larger effective root surface area. Both of these mutualistic relationships enhance nutrient uptake.

Though nitrogen is plentiful in the Earth's atmosphere, relatively few plants harbor nitrogen fixing bacteria, so most plants rely on nitrogen compounds present in the soil to support their growth. These can be supplied by mineralization of soil organic matter or added plant residues, nitrogen fixing bacteria, animal waste, or through the application offertilizers.

Hydroponics, is a method for growing plants in a water-nutrient solution without the use of nutrient-rich soil. It allows researchers and home gardeners to grow their plants in a controlled environment. The most common solution, is the Hoagland solution, developed by D. R. Hoagland in 1933, the solution consists of all the essential nutrients in the correct proportions necessary for most plant growth. An aerator is used to prevent an anoxic event or hypoxia. Hypoxia can affect nutrient uptake of a plant because without oxygen present, respiration becomes inhibited within the root cells. The Nutrient film technique is a variation of hydroponic technique. The roots are not fully submerged, which allows for adequate aeration of the roots, while a "film" thin layer of nutrient rich water is pumped through the system to provide nutrients and water to the plant.

PROCESSES

Plants take up essential elements from the soil through their roots and from the air (mainly consisting of nitrogen and oxygen) through their leaves. Nutrient uptake in the soil is achieved by cation exchange, wherein root hairs pump hydrogen ions (H+) into the soil through proton pumps. These hydrogen ions displace cations attached to negatively charged soil particles so that the cations are available for uptake by the root. In the leaves, stomata open to take in carbon dioxide and expel oxygen. The carbon dioxide molecules are used as the carbon source in photosynthesis.

FUNCTIONS OF NUTRIENTS

Although nitrogen is plentiful in the Earth's atmosphere, relatively few plants engage in nitrogen fixation (conversion of atmospheric nitrogen to a biologically useful form). Most plants therefore require nitrogen compounds to be present in the soil in which they grow.

Plant nutrition is a difficult subject to understand completely, partially because of the variation between different plants and even between different species or individuals of a given clone. Elements present at low levels may cause deficiency symptoms, and toxicity is possible at levels that are too high. Furthermore, deficiency of one element may present as symptoms of toxicity from another element, and vice versa.

Carbon and oxygen are absorbed from the air, while other nutrients are absorbed from the soil. Green plants obtain their carbohydrate supply from the carbon dioxide in the air by the process of photosynthesis. Each of these nutrients is used in a different place for a different essential function.

MACRONUTRIENTS (DERIVED FROM AIR AND WATER)

Carbon

Carbon forms the backbone of many plants biomolecules, including starches and cellulose. Carbon is fixed through photosynthesis from the carbon dioxide in the air and is a part of the carbohydrates that store energy in the plant.

Hydrogen

Hydrogen also is necessary for building sugars and building the plant. It is obtained almost entirely from water. Hydrogen ions are imperative for a proton gradient to help drive the electron transport chain in photosynthesis and for respiration.

Oxygen

Oxygen by itself or in the molecules of H_2O or CO_2 are necessary for plant cellular respiration. Cellular respiration is the process of generating energy-rich adenosine triphosphate (ATP) via the consumption of sugars made in photosynthesis. Plants produce oxygen gas during photosynthesis to produce glucose but then require oxygen to undergo aerobic cellular respiration and break down this glucose and produce ATP.

MACRONUTRIENTS (PRIMARY)

Phosphorus

Phosphorus is important in plant bioenergetics. As a component of ATP, phosphorus is needed for the conversion of light energy to chemical energy (ATP) during photosynthesis. Phosphorus can also be used to modify the activity of various enzymes by phosphorylation, and can be used for cell signaling. Since ATP can be used for thebiosynthesis of many plant biomolecules, phosphorus is important for plant growth and flower/seed formation. Phosphate esters make up DNA, RNA, and phospholipids. Most common in the form of polyprotic phosphoric acid (H_3PO_4) in soil, but it is taken up most readily in the form of H_2PO_4. Phosphorus is limited in most soils because it is released very slowly from insoluble phosphates. Under most environmental conditions it is the limiting element because of its small concentration in soil and high demand by plants and microorganisms. Plants can increase phosphorus uptake by a mutualism with mycorrhiza. A Phosphorus deficiency in plants is characterized by an intense green coloration in leaves. If the plant is experiencing high phosphorus deficiencies the leaves may become denatured and show signs of necrosis. Occasionally the leaves may appear purple from an accumulation of anthocyanin. Because phosphorus is a mobile nutrient, older leaves will show the first signs of deficiency.

It is useful to apply a high phosphorus content fertilizer, such as bone meal, to perennials to help with successful root formation.

Potassium

Potassium regulates the opening and closing of the stomata by a potassium ion pump. Since stomata are important in water regulation, potassium reduces water loss from the leaves and increases drought tolerance. Potassium deficiency may cause necrosis or interveinal chlorosis. K^+ is highly mobile and can aid in balancing the anion charges within the plant. Potassium helps in fruit colouration, shape and also increases its brix. Hence, quality fruits are produced in Potassium rich soils. It also has high solubility in water and leaches out of rocky or sandy soils. This water solubility can result in potassium deficiency. Potassium serves as an activator of enzymes used in photosynthesis and respiration Potassium is used to build cellulose and aids in photosynthesis by the formation of a chlorophyll precursor. Potassium deficiency may result in higher risk of pathogens, wilting, chlorosis, brown spotting, and higher chances of damage from frost and heat.

Nitrogen

Nitrogen is an essential component of all proteins. Nitrogen deficiency most often results in stunted growth, slow growth, and chlorosis. Nitrogen deficient plants will also exhibit a purple appearance on the stems, petioles and underside of leaves from an accumulation of anthocyanin pigments. Most of the nitrogen taken up by plants is from the soil in the forms of NO_3^-, although in acid environments such as boreal forests where nitrification is less likely to occur, ammonium NH_4^+ is more likely to be the dominating source of nitrogen. Amino acids and proteins can only be built from NH_4^+ so NO_3^- must be reduced. Under many agricultural settings, nitrogen is the limiting nutrient of high growth. Some plants require more nitrogen than others, such as corn (*Zea mays*). Because nitrogen is mobile, the older leaves exhibit chlorosis and necrosis earlier than the younger leaves. Soluble forms of nitrogen are transported as amines and amides.

MACRONUTRIENTS (SECONDARY AND TERTIARY)

Sulphur

Sulphur is a structural component of some amino acids and vitamins, and is essential in the manufacturing of chloroplasts. Sulphur is also found in the Iron Sulphur complexes of the electron transport chains in photosynthesis. It is immobile and deficiency therefore affects younger tissues first. Symptoms of deficiency include yellowing of leaves and stunted growth.

Calcium

Calcium regulates transport of other nutrients into the plant and is also involved in the activation of certain plant enzymes. Calcium deficiency results in stunting. This nutrient is involved in photosynthesis and plant structure. Blossom end rot is also a result of inadequate calcium.

Magnesium

Magnesium is an important part of chlorophyll, a critical plant pigment important in photosynthesis. It is important in the production of ATP through its role as an enzymecofactor. Magnesium deficiency can result in interveinal chlorosis.

Silicon

In plants, silicon strengthens cell walls, improving plant strength, health, and productivity. Other benefits of silicon to plants include improved drought and frost resistance, decreased lodging potential and boosting the plant's natural pest and disease fighting systems. Silicon has also been shown to improve plant vigor and physiology by improving root mass and density, and increasing above ground plant biomass and crop yields. Although not considered an essential element for plant growth and development (except for specific plant species - sugarcane and members of the horsetail family), silicon is considered a beneficial element in many countries throughout the world due to its many benefits to numerous plant species when under abiotic or biotic stresses. Silicon is currently under consideration by the Association of American Plant Food Control Officials (AAPFCO) for elevation to the status of a "plant beneficial substance".

Silicon is the second most abundant element in earth's crust. Higher plants differ characteristically in their capacity to take up silicon. Depending on their SiO_2 content they can be divided into three major groups:

- Wetland graminae-wetland rice, horsetail (10–15%)
- Dryland graminae-sugar cane, most of the cereal species and few dicotyledons species (1–3%)
- Most of dicotyledons especially legumes (<0.5%)
- The long distance transport of Si in plants is confined to the xylem. Its distribution within the shoot organ is therefore determined by transpiration rate in the organs
- The epidermal cell walls are impregnated with a film layer of silicon and effective barrier against water loss, cuticular transpiration rate in the organs.

Si can stimulate growth and yield by several indirect actions. These include decreasing mutual shading by improving leaf erectness, decreasing susceptibility to lodging, preventing Mn and Fe toxicity.

MICRO-NUTRIENTS

Some elements are directly involved in plant metabolism. However, this principle does not account for the so-called beneficial elements, whose presence, while not required, has clear positive effects on plant growth. Mineral elements that either stimulate growth but are not essential, or that are essential only for certain plant species, or under given conditions, are usually defined as beneficial elements.

Iron

Iron is necessary for photosynthesis and is present as an enzyme cofactor in plants. Iron deficiency can result in interveinal chlorosis and necrosis. Iron is not a structural part of chlorophyll but very much essential for its synthesis. Copper deficiency can be responsible for promoting an iron deficiency.

Molybdenum

Molybdenum is a cofactor to enzymes important in building amino acids. Involved in Nitrogen metabolism. Mo is part of Nitrate reductase enzyme.

Boron

Boron is important for binding of pectins in the RGII region of the primary cell wall, secondary roles may be in sugar transport, cell division, and synthesizing certain enzymes.Boron deficiency causes necrosis in young leaves and stunting.Boron is required for the uptake and utilization of calcium,membrane functioning,pollen germination,cell elongation,cell differentiation and carbohydrate metabolism.

Copper

Copper is important for photosynthesis. Symptoms for copper deficiency include chlorosis. Involved in many enzyme processes. Necessary for proper photosythesis. Involved in the manufacture of lignin (cell walls). Involved in grain production. It is also hard to find in some conditions.

Manganese

Manganese is necessary for photosynthesis, including the building of chloroplasts. Manganese deficiency may result in coloration abnormalities, such as discolored spots on the foliage.

Sodium

Sodium is involved in the regeneration of phosphoenolpyruvate in CAM and C4 plants. It can also substitute for potassium in some circumstances.

Essentiality

- Essential for C4 plants rather C3
- Substitution of K by Na: Plants can be classified into four groups:
  1. Group A—a high proportion of K can be replaced by Na and stimulate the growth, which cannot be achieved by the application of K
  2. Group B—specific growth responses to Na are observed but they are much less distinct
  3. Group C—Only minor substitution is possible and Na has no effect
  4. Group D—No substitution is occurred
- Stimulate the growth—increase leaf area, stomata, improve the water balance
- Na functions in metabolism
  1. C4 metabolism
  2. Impair the conversion of pyruvate to phosphoenol-pyruva
  3. Reduce the photosystem II activity and ultrastructural changes in mesophyll chloroplast
- Replacing K functions
  1. Internal osmoticum
  2. Stomatal function

3. Photosynthesis
4. Counteraction in long distance transport
5. Enzyme activation

- Improves the crop quality e.g. improve the taste of carrots by increasing sucrose

Zinc

Zinc is required in a large number of enzymes and plays an essential role in DNA transcription. A typical symptom of zinc deficiency is the stunted growth of leaves, commonly known as "little leaf" and is caused by the oxidative degradation of the growth hormone auxin.

Nickel

In higher plants, Nickel is absorbed by plants in the form of Ni^{2+} ion. Nickel is essential for activation of urease, an enzyme involved with nitrogen metabolism that is required to process urea. Without Nickel, toxic levels of urea accumulate, leading to the formation of necrotic lesions. In lower plants, Nickel activates several enzymes involved in a variety of processes, and can substitute for Zinc and Iron as a cofactor in some enzymes.

Chlorine

Chlorine, as compounded chloride, is necessary for osmosis and ionic balance; it also plays a role in photosynthesis.

Cobalt

Cobalt has proven to be beneficial to at least some plants, but is essential in others, such as legumes where it is required for nitrogen fixation for the symbiotic relationship it has with nitrogen-fixing bacteria. Vanadium may be required by some plants, but at very low concentrations. It may also be substituting for molybdenum. Selenium and sodium may also be beneficial. Sodium can replace potassium's regulation of stomatal opening and closing.

1. The requirement of Co for N_2 fixation in legumes and non-legumes have been documented clearly
2. Protein synthesis of Rhizobium is impaired due to Co deficiency
3. It is still not clear whether Co has direct effect on higher plant

Aluminium

- Tea has a high tolerance for Al toxicity and the growth is stimulated by Al application. The possible reason is the prevention of Cu, Mn or P toxicity effects.
- There have been reports that Al may serve as fungicide against certain types of root rot.

7

Organic Matter Recycling for Plant Nutrients

INTRODUCTION

On the basis of organic matter content, soils are characterized as mineral or organic. Mineral soils form most of the world's cultivated land and may contain from a trace to 30 percent organic matter. Organic soils are naturally rich in organic matter principally for climatic reasons. Although they contain more than 30 percent organic matter, it is precisely for this reason that they are not vital cropping soils.

This soils bulletin concentrates on the organic matter dynamics of cropping soils. In brief, it discusses circumstances that deplete organic matter and the negative outcomes of this. The bulletin then moves on to more proactive solutions. It reviews a "basket" of practices in order to show how they can increase organic matter content and discusses the land and cropping benefits that then accrue.

Soil organic matter is any material produced originally by living organisms (plant or animal) that is returned to the soil and goes through the decomposition process. At any given time, it consists of a range of materials from the intact original tissues of plants and animals to the substantially decomposed mixture of materials known as humus.

Fig. Crop residues added to the soil are decomposed by soil macrofauna and micro-organisms, increasing the organic matter content of the soil.

Most soil organic matter originates from plant tissue. Plant residues contain 60-90 percent moisture. The remaining dry matter consists of carbon (C), oxygen, hydrogen (H) and small amounts of sulphur (S), nitrogen (N), phosphorus (P), potassium (K), calcium (Ca) and magnesium (Mg). Although present in small amounts, these nutrients are very important from the viewpoint of soil fertility management.

Soil organic matter consists of a variety of components. These include, in varying proportions and many intermediate stages, an active organic fraction including microorganisms (10-40 percent), and resistant or stable organic matter (40-60 percent), also referred to as humus.

Forms and classification of soil organic matter have been described by Tate (1987) and Theng (1987). For practical purposes, organic matter may be divided into aboveground and belowground fractions. Aboveground organic matter comprises plant residues and animal residues; belowground organic matter consists of living soil fauna and microflora, partially decomposed plant and animal residues, and humic substances. The C:N ratio is also used to indicate the type of material and ease of decomposition; hard woody materials with a high C:N ratio being more resilient than soft leafy materials with a low C:N ratio.

Although soil organic matter can be partitioned conveniently into different fractions, these do not represent static end products. Instead, the amounts present reflect a dynamic equilibrium. The total amount and partitioning of organic matter in the soil is influenced by soil properties and by the quantity of annual inputs of plant and animal residues to the ecosystem. For example, in a given soil ecosystem, the rate of decomposition and accumulation of soil organic matter is determined by such soil properties as texture, pH, temperature, moisture, aeration, clay mineralogy and soil biological activities. A complication is that soil organic matter in turn influences or modifies many of these same soil properties.

Organic matter existing on the soil surface as raw plant residues helps protect the soil from the effect of rainfall, wind and sun. Removal, incorporation or burning of residues exposes the soil to negative climatic impacts, and removal or burning deprives the soil organisms of their primary energy source.

Organic matter within the soil serves several functions. From a practical agricultural standpoint, it is important for two main reasons: (i) as a "revolving nutrient fund"; and (ii) as an agent to improve soil structure, maintain tilth and minimize erosion.

As a revolving nutrient fund, organic matter serves two main functions:

- As soil organic matter is derived mainly from plant residues, it contains all of the essential plant nutrients. Therefore, accumulated organic matter is a storehouse of plant nutrients.

- The stable organic fraction (humus) adsorbs and holds nutrients in a plant-available form.

Organic matter releases nutrients in a plant-available form upon decomposition. In order to maintain this nutrient cycling system, the rate of organic matter addition from crop residues, manure and any other sources must equal the rate of decomposition, and take into account the rate of uptake by plants and losses by leaching and erosion.

Where the rate of addition is less than the rate of decomposition, soil organic matter declines. Conversely, where the rate of addition is higher than the rate of decomposition, soil organic matter increases. The term steady state describes a condition where the rate of addition is equal to the rate of decomposition.

In terms of improving soil structure, the active and some of the resistant soil organic components, together with micro-organisms (especially fungi), are involved in binding soil particles into larger aggregates. Aggregation is important for good soil structure, aeration, water infiltration and resistance to erosion and crusting.

Traditionally, soil aggregation has been linked with either total C or organic C levels. More recently, techniques have developed to fractionate C on the basis of lability (ease of oxidation), recognizing that these subpools of C may have greater effect on soil physical stability and be more sensitive indicators than total C values of carbon dynamics in agricultural systems (Lefroy, Blair and Strong, 1993; Blair, Lefroy and Lisle, 1995; Blair and Crocker, 2000). The labile carbon fraction has been shown to be an indicator of key soil chemical and physical properties. For example, this fraction has been shown to be the primary factor controlling aggregate breakdown in Ferrosols (non-cracking red clays), measured by the percentage of aggregates measuring less than 0.125 mm in the surface crust after simulated rain in the laboratory.

The resistant or stable fraction of soil organic matter contributes mainly to nutrient holding capacity (cation exchange capacity [CEC]) and soil colour. This fraction of organic matter decomposes very slowly. Therefore, it has less influence on soil fertility than the active organic fraction.

ORGANIC MATTER ORGANIC MATTER

On the basis of organic matter content, soils are characterized as mineral or organic. Mineral soils form most of the world's cultivated land and may contain from a trace to 30 percent organic matter. Organic soils are naturally rich in organic matter principally for climatic reasons. Although they contain more than 30 percent organic matter, it is precisely for this reason that they are not vital cropping soils.

This soils bulletin concentrates on the organic matter dynamics of cropping soils. In brief, it discusses circumstances that deplete organic matter and the negative outcomes of this. The bulletin then moves on to more proactive solutions. It reviews a "basket" of practices in order to show how they can increase organic matter content and discusses the land and cropping benefits that then accrue.

Soil organic matter is any material produced originally by living organisms (plant or animal) that is returned to the soil and goes through the decomposition process. At any given time, it consists of a range of materials from the intact original tissues of plants and animals to the substantially decomposed mixture of materials known as humus.

Most soil organic matter originates from plant tissue. Plant residues contain 60–90 percent moisture. The remaining dry matter consists of carbon (C), oxygen, hydrogen (H) and small amounts of sulphur (S), nitrogen (N), phosphorus (P), potassium (K), calcium (Ca) and magnesium (Mg). Although present in small amounts, these nutrients are very important from the viewpoint of soil fertility management.

Soil organic matter consists of a variety of components. These include, in varying proportions and many intermediate stages, an active organic fraction including microorganisms (10–40 percent), and resistant or stable organic matter (40–60 percent), also referred to as humus.

Forms and classification of soil organic matter have been described by Tate and Theng. For practical purposes, organic matter may be divided into aboveground and belowground fractions. Aboveground organic matter comprises plant residues and animal residues; belowground organic matter consists of living soil fauna and microflora, partially decomposed plant and animal residues, and humic substances. The C:N ratio is also used to indicate the type of material and ease of decomposition; hard woody materials with a high C:N ratio being more resilient than soft leafy materials with a low C:N ratio.

Although soil organic matter can be partitioned conveniently into different fractions, these do not represent static end products. Instead, the amounts present reflect a dynamic equilibrium. The total amount and partitioning of organic matter in the soil is influenced by soil properties and by the quantity of annual inputs of plant and animal residues to the ecosystem. For example, in a given soil ecosystem, the rate of decomposition and accumulation of soil organic matter is determined by such soil properties as texture, pH, temperature, moisture, aeration, clay mineralogy and soil biological activities. A complication is that soil organic matter in turn influences or modifies many of these same soil properties. Organic matter existing on the soil surface as raw plant residues helps protect the soil from the effect of rainfall, wind and sun. Removal, incorporation or burning of residues exposes the soil to negative climatic impacts, and removal or burning deprives the soil organisms of their primary energy source.

Organic matter within the soil serves several functions. *From a practical agricultural standpoint, it is important for two main reasons:*

(i) As a "revolving nutrient fund"; and

(ii) As an agent to improve soil structure, maintain tilth and minimize erosion.

As a revolving nutrient fund, organic matter serves two main functions:

- As soil organic matter is derived mainly from plant residues, it contains all of the essential plant nutrients. Therefore, accumulated organic matter is a storehouse of plant nutrients.
- The stable organic fraction (humus) adsorbs and holds nutrients in a plant-available form.

Organic matter releases nutrients in a plant-available form upon decomposition. In order to maintain this nutrient cycling system, the rate of organic matter addition from crop residues, manure and any other sources must equal the rate of decomposition, and take into account the rate of uptake by plants and losses by leaching and erosion.

Where the rate of addition is less than the rate of decomposition, soil organic matter declines. Conversely, where the rate of addition is higher than the rate of decomposition, soil organic matter increases. The term steady state describes a condition where the rate of addition is equal to the rate of decomposition.

In terms of improving soil structure, the active and some of the resistant soil organic components, together with micro-organisms (especially fungi), are involved in binding soil particles into larger aggregates. Aggregation is important for good soil structure, aeration, water infiltration and resistance to erosion and crusting.

Traditionally, soil aggregation has been linked with either total C or organic C levels. More recently, techniques have developed to fractionate C on the basis of lability (ease of oxidation), recognising that these subpools of C may have greater effect on soil physical stability and be more sensitive indicators than total C values of carbon dynamics in agricultural systems.

The labile carbon fraction has been shown to be an indicator of key soil chemical and physical properties. For example, this fraction has been shown to be the primary factor controlling aggregate breakdown in Ferrosols (non-cracking red clays), measured by the percentage of aggregates measuring less than 0.125 mm in the surface crust after simulated rain in the laboratory.

The resistant or stable fraction of soil organic matter contributes mainly to nutrient holding capacity (cation exchange capacity [CEC]) and soil colour. This fraction of organic matter decomposes very slowly. Therefore, it has less influence on soil fertility than the active organic fraction.

HUMAN INTERVENTIONS THAT INFLUENCE SOIL ORGANIC MATTER

Various types of human activity decrease soil organic matter contents and biological activity. However, increasing the organic matter content of soils or even maintaining good levels requires a sustained effort that includes returning organic materials to soils and rotations with high-residue crops and deep- or dense-rooting crops.

It is especially difficult to raise the organic matter content of soils that are well aerated, such as coarse sands, and soils in warm-hot and arid regions because the added materials decompose rapidly. Soil organic matter levels can be maintained with less organic residue in finetextured soils in cold temperate and moist-wet regions with restricted aeration.

PRACTICES THAT DECREASE SOIL ORGANIC MATTER

Any form of human intervention influences the activity of soil organisms and thus the equilibrium of the system. Management practices that alter the living and nutrient conditions of soil organisms, such as repetitive tillage or burning of vegetation, result in a degradation of their microenvironments.

In turn, this results in a reduction of soil biota, both in biomass and diversity. Where there are no longer organisms to decompose soil organic matter and bind soil particles, the soil structure is damaged easily by rain, wind and sun.

This can lead to rainwater run-off and soil erosion, removing the potential food for organisms, *i.e.*, the organic matter of the topsoil. Therefore, soil biota are the most important property of the soil, and "when devoid of its biota, the uppermost layer of earth ceases to be soil".

The factors leading to reduction in soil organic matter in an open cycle system can be grouped as factors that result in:

- A decrease in biomass production;
- A decrease in organic matter supply;
- Increased decomposition rates.

DECREASE IN BIOMASS PRODUCTION

Replacement of Perennial Vegetation

A consequence of clearing forest for agriculture is the disappearance of the litter layer, with a consequent reduction in the numbers and variety of soil organisms. While many temperate forest species appear to adapt well to grassland, the effects of deforestation in the tropics appear to be more marked. Studies have shown that as soil biodiversity declines, adapted species may take over from the indigenous species and the composition may change drastically.

Soil macrofaunal biomass and population density fell to 6 and 17 percent, respectively, in cultivated plots, compared with primary forest in Peruvian Amazonia. In Suriname, the number of animals per square metre has fallen to 36 percent and the diversity of species has fallen to 28 percent compared with primary forest. The indigenous species have largely disappeared, but adapted species have been available for recolonisation. The composition of the macrofaunal community has changed drastically.

DECREASE IN ORGANIC MATTER SUPPLY

Burning of Natural Vegetation and Crop Residues

The burning of maize, rice and other crop residues in the field is a common practice. Residues are usually burned to help control insects or diseases or to make fieldwork easier in the following season. Burning destroys the litter layer and so diminishes the amount of organic matter returned to the soil. The organisms that inhabit the surface soil and litter layer are also eliminated.

For future decomposition to take place, energy has to be invested first in rebuilding the microbial community before plant nutrients can be released. Similarly, fallow lands and bush are burned before cultivation. This provides a rapid supply of P to stimulate seed germination. However, the associated loss of nutrients, organic matter and soil biological activity has severe long-term consequences.

Overgrazing

There is a tendency throughout the world to overstock grazing land above its carrying capacity. Cows, draught animals and small ruminants graze on communal grazing areas and on roadsides, stream banks and other public land. Overgrazing destroys the most palatable and useful species in the plant mixture and reduces the density of the plant cover, thereby increasing the erosion hazard and reducing the nutritive value and the carrying capacity of the land.

Removal of Crop Residues

Many farmers remove residues from the field for use as animal feed and bedding or to make compost. Later, these residues return to contribute to soil fertility as manures or composts. However, residues are sometimes removed from the field and not returned. This removal of plant material impoverishes the soil as it is no longer possible to recycle the plant nutrients present in the residues.

INCREASED DECOMPOSITION RATES

Tillage Practices

Tillage is one of the major practices that reduces the organic matter level in the soil. Each time the soil is tilled, it is aerated. As the decomposition of organic matter and the liberation of C are aerobic processes, the oxygen stimulates or speeds up the action of soil microbes, which feed on organic matter.

This means that:

- When ploughed, the residues are incorporated in the soil together with air and come into contact with many micro-organisms, which accelerates the carbon cycle. The decomposition is faster, resulting in the formation of less stable humus and an increased liberation of CO_2 to the atmosphere, and thus a reduction in organic matter.
- The residues on the soil surface slow the carbon cycle because they are exposed to fewer micro-organisms and thus wane more slowly, resulting in the production of humus (which is more stable), and liberating less CO_2 to the atmosphere.

Tillage Induced Flush of Decomposition of Organic Matter.

Type of Tillage	Organic Matter Lost in 19 Days
(kg/ha) Mouldboard Plough + Disc Harrow (2x)	4300
Mouldboard Plough	2230
Disc Harrow	1840
Chisel Plough	1720
Direct Seeding	860

In terms of short-term organic matter loss, the more a soil is tilled, the more the organic matter is broken down. There are also longer-term losses, attributed to repeated, annual cultivation. Cropping systems that return little residue to the soil accelerate this decline. Many modern cropping systems combine frequent tillage with small amounts of residue, with resultant reductions in the organic matter content of many soils.

Historically, manure application (from farm livestock) was common, and it was a dynamic way of maintaining organic matter levels despite repeated cultivation and low residue returns to the soil. Increased on-farm mechanisation has reduced livestock numbers, so this source of organic material has been reduced considerably. Organic matter production and conservation is affected dramatically by conventional tillage, which not only decreases soil organic matter but also increases the potential for erosion by wind and water. *The impact occurs in many ways:*

- Ploughing leaves no residues on the soil surface to lessen the impact of rain.
- Ploughing reduces the quantity of food sources for earthworms and disturbs their burrows and living space, hence, populations of certain species decrease drastically. Moreover, reduction of earthworm numbers reduces their impact, through burrowing, in increasing porosity and aeration (particularly continuous macropores) and lowers their ability to bury and incorporate plant residues, which facilitates rapid decomposition of organic matter.
- Tillage by repeated hoeing or discing smoothes the surface and destroys natural soil aggregates and channels that connect the surface with the subsoil, leaving the soil susceptible to erosion.

Old root channels and earthworm holes are eliminated, as are the cracks between natural aggregates. The large pores, the ones destroyed by conventional tillage practices, are necessary to conduct water into the soil during rainfall.

- The development of a plough pan or hoe pan, a layer of compacted soil resulting from smearing action at the bottom of the plough or hoe, may retard both root penetration and water infiltration.
- Ploughing or discing under dry conditions exacerbates the pulverisation of the soil, causing the soil surface to crust more easily, leading to greater water run-off and erosion. This is exacerbated by reduced soil surface roughness, which leaves few depressions for temporary storage of water during intense storms.
- Increased run-off during rainstorms may also increase the possibility of drought stress later in the season, because water that runs off the field does not infiltrate into the soil to remain available to plants.

In some circumstances, imbalances of certain soil organisms can disrupt soil structure and processes, e.g., certain earthworm species in rice fields or pastures.

Drainage

Decomposition of organic matter occurs more slowly in poorly aerated soils, where oxygen is limiting or absent, compared with well-aerated soils. For this reason, organic matter accumulates in wet soil environments. Soil drainage is determined strongly by topography—soils in depressions at the bottom of hills tend to remain wet for extended periods of time because they receive water (and sediments) from upslope.

Soils may also have a layer in the subsoil that inhibits drainage, again exacerbating waterlogging and reduction in organic matter decomposition. In a permanently waterlogged soil, one of the major structural parts of plants, lignin, does not decompose at all. The ultimate consequence of extremely wet or swampy conditions is the development of organic (peat or muck) soils, with organic matter contents of more than 30 percent. Where soils are drained artificially for agricultural or other uses, the soil organic matter decomposes rapidly.

Fertilizer and Pesticide Use

Initially, the use of fertilizer and pesticides enhances crop development and thus production of biomass (especially important on depleted soils). However, the use of some fertilizers, especially N fertilizers, and pesticides can boost micro-organism activity and thus decomposition of organic matter. The chemicals provide the microorganisms with easy-to-use N components. This is especially important where the C: N ratio of the soil organic matter is high and thus decomposition is slowed by a lack of N.

PRACTICES THAT INCREASE SOIL ORGANIC MATTER

Increased concern about the environmental and economic impacts of conventional crop production has stimulated interest in alternative systems. Central to such systems is the need to promote and maintain soil biological processes and minimize fossil fuel inputs in the form of fertilizers, pesticides and mechanical cultivation. All activities aimed at the increase of organic matter in the soil help in creating a new equilibrium in the agro-ecosystem.

For a system of natural resource management to be balanced, and thus sustainable, it must be able to withstand sharp climatic fluctuations, and to evolve steadily in response to social changes and changes in the costs and availability of inputs of land, labour and knowledge. The more diverse and complex an agricultural system is the more stable and sustainable it will be in the face of unpredictable vagaries of climate and market.

Thus, annual crops, woody perennials and non-woody perennials may be combined in various ways with livestock or trees, or both, in what are now commonly called agrosilvipastoral systems.

Different approaches are required for different soil and climate conditions. However, the activities will be based on the same principle: increasing biomass production in order to build active organic matter. Active organic matter provides habitat and food for beneficial soil organisms that help build soil structure and porosity, provide nutrients to plants, and improve the water holding capacity of the soil.

Several cases have demonstrated that it is possible to restore organic matter levels in the soil. Activities that promote the accumulation and supply of organic matter, such as the use of cover crops and refraining from burning, and those that reduce decomposition rates, such as reduced and zero tillage, lead to an increase in the organic matter content in the soil.

ORGANIC MATTER DECOMPOSITION AND THE SOIL FOOD WEB

SOIL ORGANIC MATTER

When plant residues are returned to the soil, various organic compounds undergo decomposition. Decomposition is a biological process that includes the physical breakdown and biochemical transformation of complex organic molecules of dead material into simpler organic and inorganic molecules.

The continual addition of decaying plant residues to the soil surface contributes to the biological activity and the carbon cycling process in the soil. Breakdown of soil organic matter and root growth and decay also contribute to these processes. Carbon cycling is the continuous transformation of organic and inorganic carbon compounds by plants and micro- and macro-organisms between the soil, plants and the atmosphere

Decomposition of organic matter is largely a biological process that occurs naturally. Its speed is determined by three major factors: soil organisms, the physical environment and the quality of the organic matter. In the decomposition process, different products are released: carbon dioxide (CO_2), energy, water, plant nutrients and resynthesized organic carbon compounds. Successive decomposition of dead material and modified organic matter results in the formation of a more complex organic matter called humus. This process is called humification. Humus affects soil properties. As it slowly decomposes, it colours the soil darker; increases soil aggregation and aggregate stability; increases the CEC (the ability to attract and retain nutrients); and contributes N, P and other nutrients.

Soil organisms, including micro-organisms, use soil organic matter as food. As they break down the organic matter, any excess nutrients (N, P and S) are released into the soil in forms that plants can use. This release process is called mineralization. The waste products produced by micro-organisms are also soil organic matter. This waste material is less decomposable than the original plant and animal material, but it can be used by a large number of organisms. By breaking down carbon structures and rebuilding new ones or storing the C into their own biomass, soil biota plays the most important role in nutrient cycling processes and, thus, in the ability of a soil to provide the crop with sufficient nutrients to harvest a healthy product. The organic matter content, especially the more stable humus, increases the capacity to store water and store (sequester) C from the atmosphere.

THE SOIL FOOD WEB

The soil ecosystem can be defined as an interdependent life-support system composed of air, water, minerals, organic matter, and macro- and micro-organisms, all of which function together and interact closely.

The organisms and their interactions enhance many soil ecosystem functions and make up the soil food web. The energy needed for all food webs is generated by primary producers: the plants, lichens, moss, photosynthetic bacteria and algae that use sunlight to transform CO_2 from the atmosphere into carbohydrates. Most other organisms depend on the primary producers for their energy and nutrients; they are called consumers.

Soil life plays a major role in many natural processes that determine nutrient and water availability for agricultural productivity. The primary activities of all living organisms are growing and reproducing. By-products from growing roots and plant residues feed soil organisms. In turn, soil organisms support plant health as they decompose organic matter, cycle nutrients, enhance soil structure and control the populations of soil organisms, both beneficial and harmful (pests and pathogens) in terms of crop productivity.

The living part of soil organic matter includes a wide variety of micro-organisms such as bacteria, viruses, fungi, protozoa and algae. It also includes plant roots, insects, earthworms, and larger animals such as moles, mice and rabbits that spend part of their life in the soil. The living portion represents about 5 percent of the total soil organic matter. Micro-organisms, earthworms and insects help break down crop residues and manures by ingesting them and mixing them with the minerals in the soil, and in the process recycling energy and plant nutrients. Sticky substances on the skin of earthworms and those produced by fungi and bacteria help bind particles together. Earthworm casts are also more strongly aggregated (bound together) than the surrounding soil as a result of the mixing of organic matter and soil mineral material, as well as the intestinal mucus of the worm. Thus, the living part of the soil is responsible for keeping air and water available, providing plant nutrients, breaking down pollutants and maintaining the soil structure.

The composition of soil organisms depends on the food source (which in turn is season dependent). Therefore, the organisms are neither uniformly distributed through the soil nor uniformly present all year. However, in some cases their biogenic structures remain. Each species and group exists where it can find appropriate food supply, space, nutrients and moisture. Organisms occur wherever organic matter occurs. Therefore, soil organisms are concentrated: around roots, in litter, on humus, on the surface of soil aggregates and in spaces between aggregates. For this reason, they are most prevalent in forested areas and cropping systems that leave a lot of biomass on the surface.

The activity of soil organisms follows seasonal as well as daily patterns. Not all organisms are active at the same time. Most are barely active or even dormant. Availability of food is an important factor that influences the level of activity of soil organisms and thus is related to land use and management. Practices that increase numbers and activity of soil organisms include: no tillage or minimal tillage; and the maintenance of plant and annual residues that reduce disturbance of soil organisms and their habitat and provide a food supply.

Different groups of organisms can be distinguished in the soil.

DECOMPOSITION PROCESS

Fresh residues consist of recently deceased micro-organisms, insects and earthworms, old plant roots, crop residues, and recently added manures.

Crop residues contain mainly complex carbon compounds originating from cell walls (cellulose, hemicellulose, etc.). Chains of carbon, with each carbon atom linked to other carbons, form the "backbone" of organic molecules. These carbon chains, with varying amounts of attached oxygen,

H, N, P and S, are the basis for both simple sugars and amino acids and more complicated molecules of long carbon chains or rings. Depending on their chemical structure, decomposition is rapid (sugars, starches and proteins), slow (cellulose, fats, waxes and resins) or very slow (lignin).

Table. Classification of soil organisms

Micro-organisms	**Microflora**	**<5 μm**	**Bacteria Fungi**
	Microfauna	<100 μm	Protozoa Nematodes
Macro-organisms	Meso-organisms	100 μm - 2 mm	Springtails Mites
	Macro-organisms	2 - 20 mm	Earthworms Millipedes Woodlice Snails and slugs
Plants	Algae	10 μm	
	Roots	> 10 μm	

Note:

Clay particles are smaller than 2 μm.

Table. Essential functions performed by different members of soil organisms (biota)

Functions	**Organisms involved**
Maintenance of soil structure	Bioturbating invertebrates and plant roots, mycorrhizae and some other micro-organisms
Regulation of soil hydrological processes	Most bioturbating invertebrates and plant roots
Gas exchange and carbon sequestration (accumulation in soil)	Mostly micro-organisms and plant roots, some C protected in large compact biogenic invertebrate aggregates
Soil detoxification	Mostly micro-organisms
Nutrient cycling	Mostly micro-organisms and plant roots, some soil- and litter-feeding invertebrates
Decomposition of organic matter	Various saprophytic and litter-feeding invertebrates (detritivores), fungi, bacteria, actinomycetes and other micro-organisms
Suppression of pests, parasites and diseases	Plants, mycorrhizae and other fungi, nematodes, bacteria and various other micro-organisms, collembola, earthworms, various predators
Sources of food and medicines	Plant roots, various insects (crickets, beetle larvae, ants, termites), earthworms, vertebrates, micro-organisms and their by-products
Symbiotic and asymbiotic relationships with plants and their roots	Rhizobia, mycorrhizae, actinomycetes, diazotrophic bacteria and various other rhizosphere micro-organisms, ants
Plant growth control (positive and negative)	Direct effects: plant roots, rhizobia, mycorrhizae, actinomycetes, pathogens, phytoparasitic nematodes, rhizophagous insects, plant-growth promoting rhizosphere micro-organisms, biocontrol agents Indirect effects: most soil biota

During the decomposition process, microorganisms convert the carbon structures of fresh residues into transformed carbon products in the soil. There are many different types of organic molecules in soil. Some are simple molecules that have been synthesized directly from plants or other living organisms. These relatively simple chemicals, such as sugars, amino acids, and cellulose are readily consumed by many organisms. For this reason, they do not remain in the soil for a long time. Other chemicals such as resins and waxes also come directly from plants, but are more difficult for soil organisms to break down.

Humus is the result of successive steps in the decomposition of organic matter. Because of the complex structure of humic substances, humus cannot be used by many micro-organisms as an energy source and remains in the soil for a relatively long time.

Non-humic substances: significance and function

Non-humic organic molecules are released directly from cells of fresh residues, such as proteins, amino acids, sugars, and starches. This part of soil organic matter is the active, or easily decomposed, fraction. This active fraction is influenced strongly by weather conditions, moisture status of the soil, growth stage of the vegetation, addition of organic residues, and cultural practices, such as tillage. It is the main food supply for various organisms in the soil. Carbohydrates occur in the soil in three main forms: free sugars in the soil solution, cellulose and hemicellulose; complex polysaccharides; and polymeric molecules of various sizes and shapes that are attached strongly to clay colloids and humic substances. The simple sugars, cellulose and hemicellulose, may constitute 5-25 percent of the organic matter in most soils, but are easily broken down by micro-organisms.

Polysaccharides (repeating units of sugar-type molecules connected in longer chains) promote better soil structure through their ability to bind inorganic soil particles into stable aggregates. Research indicates that the heavier polysaccharide molecules may be more important in promoting aggregate stability and water infiltration than the lighter molecules. Some sugars may stimulate seed germination and root elongation. Other soil properties affected by polysaccharides include CEC, anion retention and biological activity.

The soil lipids form a very diverse group of materials, of which fats, waxes and resins make up 2-6 percent of soil organic matter. The significance of lipids arises from the ability of some compounds to act as growth hormones. Others may have a depressing effect on plant growth.

Soil N occurs mainly (> 90 percent) in organic forms as amino acids, nucleic acids and amino sugars. Small amounts exist in the form of amines, vitamins, pesticides and their degradation products, etc. The rest is present as ammonium (NH_4^-) and is held by the clay minerals.

Compounds and function of humus

Humus or humified organic matter is the remaining part of organic matter that has been used and transformed by many different soil organisms.

It is a relatively stable component formed by humic substances, including humic acids, fulvic acids, hymatomelanic acids and humins. It is probably the most widely distributed organic carbon-containing material in terrestrial and aquatic environments. Humus cannot be decomposed readily because of its intimate interactions with soil mineral phases and is chemically too complex to be used by most organisms. It has many functions.

One of the most striking characteristics of humic substances is their ability to interact with metal ions, oxides, hydroxides, mineral and organic compounds, including toxic pollutants, to form water-soluble and water-insoluble complexes. Through the formation of these complexes, humic substances can dissolve, mobilize and transport metals and organics in soils and waters, or accumulate in certain soil horizons. This influences nutrient availability, especially those nutrients present at microconcentrations only. Accumulation of such complexes can contribute to a reduction of toxicity, e.g. of aluminium (Al) in acid soils, or the capture of pollutants - herbicides such as Atrazine or pesticides such as Tefluthrin - in the cavities of the humic substances.

Humic and fulvic substances enhance plant growth directly through physiological and nutritional effects. Some of these substances function as natural plant hormones (auxines and gibberillins) and are capable of improving seed germination, root initiation, uptake of plant nutrients and can serve as sources of N, P and S. Indirectly, they may affect plant growth through modifications of physical, chemical and biological properties of the soil, for example, enhanced soil water holding capacity and CEC, and improved tilth and aeration through good soil structure.

About 35-55 percent of the non-living part of organic matter is humus. It is an important buffer, reducing fluctuations in soil acidity and nutrient availability. Compared with simple organic molecules, humic substances are very complex and large, with high molecular weights. The characteristics of the well-decomposed part of the organic matter, the humus, are very different from those of simple organic molecules. While much is known about their general chemical composition, the relative significance of the various types of humic materials to plant growth is yet to be established.

Humus consists of different humic substances:

- Fulvic acids: the fraction of humus that is soluble in water under all pH conditions. Their colour is commonly light yellow to yellow-brown.
- Humic acids: the fraction of humus that is soluble in water, except for conditions more acid than pH 2. Common colours are dark brown to black.
- Humin: the fraction of humus that is not soluble in water at any pH and that cannot be extracted with a strong base, such as sodium hydroxide (NaOH). Commonly black in colour.

The term acid is used to describe humic materials because humus behaves like weak acids.

Fulvic and humic acids are complex mixtures of large molecules. Humic acids are larger than fulvic acids. Research suggests that the different substances are differentiated from each other on the basis of their water solubility.

Fulvic acids are produced in the earlier stages of humus formation. The relative amounts of humic and fulvic acids in soils vary with soil type and management practices. The humus of forest soils is characterized by a high content of fulvic acids, while the humus of agricultural and grassland areas contains more humic acids.

NATURAL FACTORS INFLUENCING THE AMOUNT OF ORGANIC MATTER

The transformation and movement of materials within soil organic matter pools is a dynamic process influenced by climate, soil type, vegetation and soil organisms. All these factors operate within a hierarchical spatial scale. Soil organisms are responsible for the decay and cycling of both macronutrients and micronutrients, and their activity affects the structure, tilth and productivity of the soil.

In natural humid and subhumid forest ecosystems without human disturbance, the living and non-living components are in dynamic equilibrium with each other. The litter on the soil surface beneath different canopy layers and high biomass production generally result in high biological activity in the soil and on the soil surface. Mollison and Slay (1991) distinguished the following five mechanisms:

- A continuous soil cover of living plants, which together with the soil architecture facilitates the capture and infiltration of rainwater and protects the soil;
- A litter layer of decomposing leaves or residues providing a continuous energy source for macro- and micro-organisms;
- The roots of different plants distributed throughout the soil at different depths permit an effective uptake of nutrients and an active interaction with microorganisms;
- The major period of nutrient release by micro-organisms coincides with the major period of nutrient demand by plants;
- Nutrients recycled by deep-rooting plants and soil macrofauna and microfauna.

This equilibrium creates almost closed-cycle transfers of nutrients between soil and the vegetation adapted to such site conditions, resulting in almost perfect physical and hydric conditions for plant growth, i.e. a cool microclimate, increased evapotranspiration, good rooting conditions with good porosity and sufficient soil moisture. This facilitates water infiltration and prevents erosion and runoff. Thus, it results in clean water in the streams emanating from the area, a relatively smooth variation in streamflow during the year, and recharge of groundwater.

In human-managed systems, the soil biological activity is influenced by the land use system, plant types and the management practices. The influence of land management practices. The environmental and edaphic factors that control the activity of soil biota, and thus the balance between accumulation and decomposition of organic matter in the soil.

TEMPERATURE

Several field studies have shown that temperature is a key factor controlling the rate of decomposition of plant residues. Decomposition normally occurs more rapidly in the tropics than in temperate areas. Ladd and Amato (1985) reported that, despite differences in plant material and climate patterns, the decomposition of leguminous materials in southern Australian sites followed the same pattern as that of ryegrass for sites in Nigeria and the United Kingdom, although the time scales were different. Reaction rates doubled for each increase of 8-9 °C in the mean annual air temperature. The relatively faster rate of decomposition induced by the continuous warmth in the tropics implies that high equilibrium levels of organic matter are difficult to achieve in tropical agro-ecosystems. Hence, large annual rates of organic inputs are needed to maintain an adequate labile soil organic matter pool in cultivated soils. Soils in cooler climates commonly have more organic matter because of slower mineralization (decomposition) rates.

SOIL MOISTURE AND WATER SATURATION

Soil organic matter levels commonly increase as mean annual precipitation increases. Conditions of elevated levels of soil moisture result in greater biomass production, which provides more residues, and thus more potential food for soil biota.

Soil biological activity requires air and moisture. Optimal microbial activity occurs at near “field capacity”, which is equivalent to 60-percent water-filled pore space.

On the other hand, periods of water saturation lead to poor aeration. Most soil organisms need oxygen, and thus a reduction of oxygen in the soil leads to a reduction of the mineralization rate as these organisms become inactive or even die. Some of the transformation processes become anaerobic, which can lead to damage to plant roots caused by waste products or favourable conditions for disease-causing organisms. Continued production and slow decomposition can lead to very large organic matter contents in soils with long periods of water saturation (e.g. peat soils, and tea crops in India).

With the exception of the hyperhumid regions, the climates of vast areas of the humid, subhumid and semi-arid tropics are characterized by distinct wet and dry seasons. In the wet-dry tropics, large amounts of nitrate often occur in the surface soil during the first part of the rainy season. This accelerated nitrogen mineralization caused by a large increase in microbial activity is the result of the first few rains activating the labile soil organic matter.

Farmers who practise "slash and burn" agriculture often choose early planting in order to take advantage of this flush of inorganic N before it is lost through leaching and runoff. In these low-input systems, the amount of nitrate present in the soil during the early part of the rainy season is related closely to the organic matter content of the soil. N availability diminishes during the later part of the rainy season.

SOIL TEXTURE

Soil organic matter tends to increase as the clay content increases. This increase depends on two mechanisms. First, bonds between the surface of clay particles and organic matter retard the decomposition process. Second, soils with higher clay content increase the potential for aggregate formation. Macroaggregates physically protect organic matter molecules from further mineralization caused by microbial attack. For example, when earthworm casts and the large soil particles they contain are split by the joint action of several factors (climate, plant growth and other organisms), nutrients are released and made available to other components of soil micro-organisms.

Under similar climate conditions, the organic matter content in fine textured (clayey) soils is two to four times that of coarse textured (sandy) soils.

Kaolinite, the main clay mineral in many upland soils in the tropics, has a much smaller specific surface and nutrient exchange capacity than most other clay minerals. Therefore, kaolinitic soils contain considerably fewer clay-humus complexes. In addition, the unprotected labile humic substances are vulnerable to decomposition under appropriate soil moisture conditions. Thus, high levels of organic matter are difficult to maintain in cultivated kaolinitic soils in the wet-dry tropics, because climate and soil conditions favour rapid decomposition. In contrast, organic matter can persist as organo-oxide complexes in soils rich in iron and aluminium oxides. Such properties favour the formation of soil microaggregates, typical of many fine-textured, oxide-rich, high base-status soils in the tropics. These soils are known for their low bulk density, high microporosity, and high organic-matter retention under natural vegetation, but also for their high phosphate fixation capacity on the oxides when used for crop production. Current knowledge suggests that whereas organic matter contributes to the dark colour of Vertisols, it is not considered important in determining either the development, robustness or resilience of structure in these soils. Organic matter levels tend to be low in Vertisols; even as low as 10 g/ kg.

Parent material influences organic matter accumulation not only through its effect on soil texture. Soils developed from inherently rich material, such as basalt, are more fertile than soils formed from granitic material, which contains less mineral nutrients. Moreover, the former experience more organic matter accumulation because of abundant vegetative growth.

TOPOGRAPHY

Organic matter accumulation is often favoured at the bottom of hills. There are two reasons for this accumulation: conditions are wetter than at mid- or upper-slope positions, and organic matter is transported to the lowest point in the landscape through runoff and erosion. Similarly, soil organic matter levels are higher on northfacing slopes (in the Northern Hemisphere) compared with south-facing slopes (and the other way around in the Southern Hemisphere) because temperatures are lower.

SALINITY AND ACIDITY

Salinity, toxicity and extremes in soil pH (acid or alkaline) result in poor biomass production and, thus in reduced additions of organic matter to the soil. For example, pH affects humus formation in two ways: decomposition, and biomass production. In strongly acid or highly alkaline soils, the growing conditions for micro-organisms are poor, resulting in low levels of biological oxidation of organic matter. Soil acidity also influences the availability of plant nutrients and thus regulates indirectly biomass production and the available food for soil biota. Fungi are less sensitive than bacteria to acid soil conditions.

VEGETATION AND BIOMASS PRODUCTION

The rate of soil organic matter accumulation depends largely on the quantity and quality of organic matter input. Under tropical conditions, applications of readily degradable materials with low C:N ratios, such as green manure and leguminous cover crops, favour decomposition and a short-term increase in the labile nitrogen pool during the growing season. On the other hand, applications of plant materials with both large C:N ratios and lignin contents such as cereal straw and grasses generally favour nutrient immobilization, organic matter accumulation and humus formation, with increased potential for improved soil structure development.

Plant constituents such as lignin and other polyphenols retard decomposition. In an experiment in southern Nigeria to compare management effects on soil organic matter accumulation, a three-year fallow with Guinea grass (*Panicum maximum*), which has a high lignin content, maintained a carbon level comparable to that under forest fallow. However, fallowing with leguminous species such as pigeon pea (*Cajanus cajan*) caused a significant decline in soil total C.

Palm and Sanchez (1990) reported that both the decomposition rate and the N-release patterns of three tropical legumes (*Inga edulis, Cajanus cajan*, and *Erythrina* spp.) were related to the amount of polyphenol compounds such as lignin in the leaf. *Erythrina* leaves had the lowest concentrations of polyphenols and the fastest decomposition rate of the three species studied. Root turnover also constitutes an important addition of humus into the soil, and consequently it is important for carbon sequestration. In forests, most organic matter is added as superficial litter. However, in grassland ecosystems, up to two-thirds of organic matter is added through the decay of roots.

NUTRIENT CYCLING AND SOIL ORGANIC MATTER

In a soil-plant system, plant nutrients are in a state of continuous, dynamic transfer. The plants take up the nutrients from the soil and use them for metabolic processes. Some of the plant parts such as dead leaves and roots are returned to the soil during the plant's growth, and, depending upon the type of land use and the nature of plants, plant parts are added to the soil when the plants are harvested. The litter or biomass so added decomposes through the activity of soil microorganisms, and the nutrients that had been bound in the plant parts are released to the soil where they become once again available to be taken up by plants. In a limited sense, nutrient cycling refers to this continuous transfer of nutrients from soil to plant and back to soil, and this is the sense in which it is used in this chapter. In a broader sense, nutrient cycling involves the continuous transfer of nutrients within different components of the ecosystem and includes processes such as weathering of minerals, activities of soil biota, and other transformations occurring in the biosphere, atmosphere, lithosphere, and hydrosphere.

Nutrient cycling occurs to varying degrees in all land-use systems. Agroforestry and other tree-based systems are commonly credited with more efficient nutrient cycling (and, in turn, a greater potential to improve soil fertility) than many other systems because of the presence of woody perennials in the system and their suggested beneficial effects on the soil. These woody perennials have, theoretically, more extensive and deeper root systems than herbaceous plants and thus have a potential to capture and recycle a larger amount of nutrients. Their litter contribution to the soil's surface is probably also greater than that of herbaceous plants.

NUTRIENT CYCLING IN TROPICAL IN TROPICAL FOREST ECOSYSTEMS

A simplified model of the nutrient cycle in a forest ecosystem. The model consists of the soil-plant system which is partitioned into several compartments. The crown surface forms the boundary of the system; this is where input of bioelements (i.e., elements that are biologically important) occurs through precipitation. The soil surface is the entry point for inputs into the soil compartment. The surface soil layer is considered as the zone of intensive root-activity, whereas the subsoil constitutes the extensive root-activity zone. The deeper limits of the extensive root layer is the boundary between the ecosystem and the hydrosphere and lithosphere. Bioelements transported beyond this layer are lost from the ecosystem and appear as output from the system.

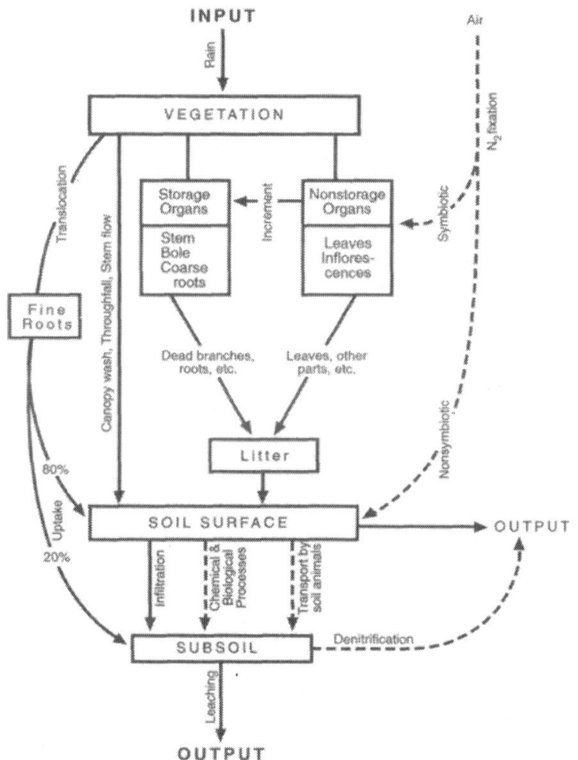

Fig. A simplified model of nutrient cycling in a forest ecosystem. Dotted lines indicate the biological processes taking place with the involvement of other organisms.

Nutrients taken up by the plant are either stored in an increment (storage) compartment or are used for the production of nonstorage organs. It is well known that fertilizer application initially results in an input of nutrients and accompanying ions into the solid phase of the uppermost layer of the soil. Depending on the water content of the soil and the solubility of the fertilizers, they pass into the solution phase of the same soil layer, and then spread -depending on the mobility of the ions in different layers of the soil and many other factors - into the plant stand via uptake. Based on the flow rate of percolating water and the soil properties, part of the nutrients that are in the soil solution are washed out of the nutrient-absorbing zone, and this represents a loss (output) from the system. Dissolved nutrients, especially ions like nitrate, which do not significantly interact with the soil matrix (i.e., are not "held" by it), have a greater likelihood of being lost in unsaturated water flow. Phosphates which possess low solubilities or are transformed into compounds of low solubility are least affected by leaching or percolation loss, whereas the magnitude of loss of cations like potassium depends on the exchange capacity of soils.

Some of the nutrients that are taken up by the plants are subsequently returned to the soil through two avenues. First, through litterfall and secondly, through the process of plant cycling. The latter represents that part of the total uptake of nutrients which is leached out of the vegetative canopy and returned to the soil via crown drip and stem flow. Although this phenomenon is usually not recognized in nutrient cycling studies, its contribution is important, especially in deciduous trees. The presence of this fraction, which is circulated within the ecosystem, is a sort of "necessary waste", and it should be recognized as part of the nutrient loss that must be accounted for while calculating nutrient budgets of plant communities. The total amount involved in this cycling depends on several factors such as the nutrient content of the leaves, intensity and frequency of rain, and age and arrangement of leaves.

Ecosystems composed predominantly of trees characteristically contain large quantities of living biomass (including wood) and therefore, large accumulations of chemical elements. About 20 to 30% of the total living biomass of the trees is in their roots, and there is a constant addition of organic matter to the soil through dead and decaying roots. Nye (1961) estimated that under a moist tropical forest the net annual contribution of dead roots was around 2,600 kg ha^{-1}. In addition, there can be significant additions of soil organic matter during active root growth in the form of sloughed-off tissue, much of it coming directly from the roots without the intervention of soil microfauna. In tracer experiments with annual plants (wheat and mustard) Sauerbeck and Johnen (1977) found that the total deposition of photosynthate in roots during the growth period was 2 to 3 times greater than the total quantity of roots present at the end of the growth period. Martin (1977) suggested that this was caused by sloughing-off of root tissues during the active growth of the plant, and it represented a steady release of carbohydrate-rich organic material from actively growing roots, and, thus, an energy input into the soil ecosystem capable of supporting a substantial microbial population. This phenomenon would be especially important in the organic matter and nutrient relations of soils under trees.

The deep-rooting characteristics of most trees are often cited as being desirable for agroforestry systems. The basis of this assumption is that, because of their deep roots, trees are able to absorb nutrients from soil depths that crop roots cannot reach. However, data are needed to substantiate this. Most of the fine, feeder roots of many common trees are found within the 20 cm-deep topsoil. Radio-tracer techniques have been used extensively in studies of the root systems of tree crops, such as cacao, apple, coffee, and guava ; but most of these studies have focused on the extent of the root systems, rather than on variations in uptake according to different soil depths. These studies have also shown that although subsoil nutrients can play an important role in orchard tree nutrition, nutrient uptake is not directly proportional to root volume.

The major recognized avenue for addition of organic matter to the soil (and, hence, of nutrients to the soil from the trees standing on it), is through litter fall, that is, through dead and falling leaves, twigs, branches, fruits, and so on. There are several studies on this process in tropical forests, which include Malaisse et al. (1975) in Africa; Kira and Shidei (1967), and Kira (1969) in Asia; Klinge and Rodrigues (1968), Medina (1968), Cornforth (1970), Klinge (1977) and Kunkel-Westphal and Kunkel (1979) in South America; and Edwards (1977) in New Guinea. However, the results of these nutrient cycling studies in forest ecosystems may not be of direct relevance to agroforestry systems, because compared to forest ecosystems, agroforestry systems are subject to more frequent disturbances caused by management practices such as pruning and soil tillage. Some data on nutrient addition to the soil via litterfall/prunings in various agroforestry systems in the humid tropics are given inTable 16.1. As could be expected, there is considerable variation in the data because of the differences in site characteristics, management practices, methods of sampling and analyses, etc.

The major avenue of output or removal of nutrients from a managed system is export through harvested produce. Such exports are generally greater for annual agricultural crops in terms of the total quantity removed per unit area and unit time. In the case of woody perennials, it depends on the frequency and intensity of harvesting. However, because even repeated harvests of seasonal products such as fruits, leaves, and latex do not amount to destructive or total harvesting in woody perennials, the rates of their export out of the soil-plant system are relatively low compared to annual agricultural crops.

NUTRIENT CYCLING IN AGROFORESTRY SYSTEMS

A schematic presentation of the general pattern of nutrient cycling in an agroforestry system in comparison with an agricultural system and a forestry system. It should be noted that the cycles for nitrogen, phosphorus, potassium, and other elements vary considerably, and should be considered separately. However, they all have some common characteristics as indicated in the model. The cycle consists of inputs into (gains), output from (losses), and internal turnover or transfer within the system as depicted in the forestry. The paths of these gains, losses, and transfers are also similar: inputs come through fertilizer, rain, dust, organic materials from outside the system, and N_2 fixation (for N) as well as weathering of rocks (for other elements); the principal outputs are derived from erosion, percolation (leaching), and crop harvest (for all nutrients), denitrification and volatilization (for N), and burning (for N and S).

Forest ecosystems represent closed and efficient nutrient cycling systems, meaning that they have high rates of turnover, and low rates of outputs or losses from (as well as inputs into) the system; in other words, they are self-sustaining. On the other hand, common agricultural systems are often open or "leaky," meaning that the turnover within the system is relatively low and losses as well as inputs are comparatively high. Nutrient cycling in agroforestry systems falls between these two extremes; more nutrients in the system are re-used by plants (compared to agricultural systems) before being lost from the system. The major difference between agroforestry and other land-use systems lies in the transfer or turn-over of nutrients within the system from one component to the other, and the possibility of managing the system or its components to facilitate increased rates of turn-over without affecting the overall productivity of the system. Results from a number of studies support this view.

Sanchez (1987), in a review of this topic, cited encouraging results from experiments conducted to assess the nutrient cycling potential of agroforestry systems on Alfisols and Andepts of moderate to high fertility. Studies on the use of *Erythrina poeppigiana* as shade trees in *Coffea arabica* plantations in Costa Rica have also yielded promising results. Juo and Lai (1977) compared the effects of *Leucaena leucocephala* fallow versus a bush fallow on selected chemical properties of an Alfisol in western Nigeria. After three years, during which *L. leucocephala* biomass was cut annually and returned to the soil as mulch, the cation exchange capacity and levels of exchangeable calcium and potassium were significantly higher in the *L. leucocephala*fallow than in the bush fallow. However, MacDicken (1991) reported from his studies in Occidental Mindoro, The Philippines, that soil pH in the 0-10 cm depth was significantly higher under a natural bush fallow compared with an improved fallow planted with *L. leucocephala;* he attributed this to increased extraction of Ca from lower soil depths and its deposition on the upper layers under the bush fallow. The results of several years of investigations on the acid soils (Ultisols: Typic Paleudults) of Yurimaguas in the Amazon basin of Peru showed that managed leguminous fallows significantly increased soil nutrient (N and P) levels. Studies carried out by

Agamuthu and Broughton (1985) showed that nutrient cycling in oil-palm plantations where leguminous cover crops (*Centrosema pubescens* and *Pueraria phaseoloides*) were used was more efficient than in plantations where there was no cover crop. Besides the addition of about 150 kg N ha^{-1} yr^{-1}, the loss of nitrate nitrogen through leaching was significantly lower in the former system. This indicates that the improved cycling is perhaps due to the presence of the leguminous species - no matter whether it is woody or nonwoody - and not of the woody perennial, *per se*.

Young (1989), reviewing studies on the nitrogen content of litterfall and prunings, provided data on various tree species in agroforestry systems in humid and moist subhumid climates, and compared these with data from natural vegetation communities. In alley-cropping systems, some species are capable of supplying 100-200 kg N ha^{-1} yr^{-1} (or, even more) if all the prunings are left on the soil; this is approximately the same as the amount of nitrogen that is removed during harvest in cereal/legume intercropping systems. In coffee and cacao plantations with shade trees (some of which are N_2 fixing), the return from litter and prunings is 100-300 kg N ha^{-1} yr^{-1}, which is much higher than the amount removed during harvest or derived from nitrogen fixation.

A number of studies on soil changes under shifting cultivation have been conducted (Jordan et al., 1983; Toky and Ramakrishnan, 1983; Andriesse and Koopmans, 1984; Andriesse and Schelhaas, 1985). However, there are no data as yet on nutrient cycling in agroforestry systems based on shifting cultivation. The major drawback with respect to nutrients in shifting cultivation is that most of the nutrients built up during the fallow period is in the vegetation, and some of it, especially nitrogen, is lost when the vegetation is burned.

Some tree and shrub species can selectively accumulate certain nutrients, even in soils which contain very small amounts of these nutrients. Palms, for example, are able to accumulate large amounts of potassium, tree ferns accumulate nitrogen, *Gmelina arborea* accumulates calcium and *Cecropia species* growing on acid soils appear to accumulate calcium and phosphorus. However, as Golley (1986) cautioned, the ability to accumulate nutrients varies according to particular sites and soils, and this factor must be taken into account while selecting nutrient-conserving species for incorporation into agroforestry technologies. There is also some evidence that indicates a higher nutrient content in trees or bushes that are scattered over extensive areas. The extent and mechanism of soil nutrient enrichment by some of the common savanna tree and shrub species growing in highly weathered and infertile Ultisols of the Mountain Pine Ridge savannas of Belize (17°N latitude, 89°W longitude), Kellman (1979) reported that trees enriched the soil below them in terms of Ca, Mg, K, Na, P and N. In some cases the levels of these nutrients approached or exceeded those found in the nearby rainforest.

NUTRIENT CYCLE

Fig. Composting within agricultural systems capitalizes upon the natural services of nutrient recycling in ecosystems. Bacteria,fungi, insects, earthworms, bugs, and other creatures dig and digest the compost into fertile soil. The minerals and nutrients in the soil is recycled back into the production of crops.

A nutrient cycle (or ecological recycling) is the movement and exchange of organic and inorganic matter back into the production of living matter. The process is regulated by food web pathways that decompose matter intomineral nutrients. Nutrient cycles occur within ecosystems. Ecosystems are interconnected systems where matter and energy flows and is exchanged as organisms feed, digest, and migrate about. Minerals and nutrients accumulate in varied densities and uneven configurations across the planet. Ecosystems recycle locally, converting mineral nutrients into the production of biomass, and on a larger scale they participate in a global system of inputs and outputs where matter is exchanged and transported through a larger system of biogeochemical cycles.

Particulate matter is recycled by biodiversity inhabiting the detritus in soils, water columns, and along particle surfaces (including 'aeolian dust'). Ecologists may refer to ecological recycling, organic recycling, biocycling, cycling, biogeochemical recycling, natural recycling, or just recycling in reference to the work of nature. Whereas the global biogeochemical cycles describe the natural movement and exchange of every kind of particulate matter through the living and non-living components of the Earth, nutrient cycling refers to the biodiversity within community food web systems that loop organic nutrients or water supplies back into production. The difference is a matter of scale and compartmentalization with nutrient cycles feeding into global biogeochemical cycles. Solar energy flows through ecosystems along unidirectional and noncyclic pathways, whereas the movement of mineral nutrients is cyclic. Mineral cycles include carbon cycle, sulfur cycle, nitrogen cycle, water cycle, phosphorus cycle,oxygen cycle, among others that continually recycle along with other mineral nutrients into productive ecological nutrition. Global biogeochemical cycles are the sum product of localized ecological recycling regulated by the action of food webs moving particulate matter from one living generation onto the next. Earths ecosystems have recycled mineral nutrients sustainably for billions of years.

OUTLINE

Fig. Fallen logs are critical components of the nutrient cycle in terrestrial forests. Nurse logs form habitats for other creatures that decompose the materials and recycle the nutrients back into production.

The nutrient cycle is nature's recycling system. All forms of recycling have feedback loops that use energy in the process of putting material resources back into use. Recycling in ecology is regulated to a large extent during the process of decomposition. Ecosystems employ biodiversity in the food webs that recycle natural materials, such as mineral nutrients, which includes water. Recycling in natural systems is one of the many ecosystem services that sustain and contribute to the well-being of human societies.

There is much overlap between the terms for biogeochemical cycle and nutrient cycle. Most textbooks integrate the two and seem to treat them as synonymous terms. However, the terms often appear independently. Nutrient cycle is more often used in direct reference to the idea of an intra-system cycle, where an ecosystem functions as a unit. From a practical point it does not make sense to assess a terrestrial ecosystem by considering the full column of air above it as well as the great depths of Earth below it. While an ecosystem often has no clear boundary, as a working model it is practical to consider the functional community where the bulk of matter and energy transfer occurs. Nutrient cycling occurs in ecosystems that participate in the "larger biogeochemical cycles of the earth through a system of inputs and outputs."

COMPLETE AND CLOSED LOOP

All systems recycle. The biosphere is a network of continually recycling materials and information in alternating cycles of convergence and divergence. As materials converge or become more concentrated they gain in quality, increasing their potentials to drive useful work in proportion to their concentrations relative to the environment. As their potentials are used, materials diverge, or become more dispersed in the landscape, only to be concentrated again at another time and place.

Ecosystems are capable of complete recycling. Complete recycling means that 100% of the waste material can be reconstituted indefinitely. This idea was captured by Howard T. Odumwhen he penned that "it is thoroughly demonstrated by ecological systems and geological systems that all the chemical elements and many organic substances can be accumulated by living systems from background crustal or oceanic concentrations without limit as to concentration so long as there is available solar or other source of potential energy" In 1979 Nicholas Georgescu-Roegen proposed a fourth law of entropy stating that complete recycling is impossible. Despite Georgescu-Roegen's extensive intellectual contributions to the science of ecological economics, the fourth law has been rejected in line with observations of ecological recycling. However, some authors state that complete recycling is impossible for technological waste.

Ecosystems execute closed loop recycling where demand for the nutrients that adds to the growth of biomass exceeds supply within that system. There are regional and spatial differences in the rates of growth and exchange of materials, where some ecosystems may be in nutrient debt (sinks) where others will have extra supply (sources). These differences relate to climate, topography, and geological history leaving behind different sources of parent material. In terms of a food web, a cycle or loop is defined as "a directed sequence of one or more links starting from, and ending at, the same species." An example of this is the microbial food web in the ocean, where "bacteria are exploited, and controlled, by protozoa, including heterotrophic microflagellates which are in turn exploited by ciliates. This grazing activity is accompanied by excretion of substances which are in turn used by the bacteria, so that the system more or less operates in a closed circuit."

ECOLOGICAL RECYCLING

A large fraction of the elements composing living matter reside at any instant of time in the world's biota. Because the earthly pool of these elements is limited and the rates of exchange among the various components of the biota are extremely fast with respect to geological time, it is quite evident that much of the same material is being incorporated again and again into different biological forms. This observation gives rise to the notion that, on the average, matter (and some amounts of energy) are involved in cycles.

An example of ecological recycling occurs in the enzymaticdigestion of cellulose. "Cellulose, one of the most abundant organic compounds on Earth, is the major polysaccharide in plants where it is part of the cell walls. Cellulose-degrading enzymes participate in the natural, *ecological recycling* of plant material." Different ecosystems can vary in their recycling rates of litter, which creates a complex feedback on factors such as the competitive dominance of certain plant species. Different rates and patterns of ecological recycling leaves a legacy of environmental effects with implications for the future evolution of ecosystems.

Ecological recycling is common in organic farming, where nutrient management is *fundamentally different*compared to agri-business styles of soil management. Organic farms that employ ecosystem recycling to a greater extent support more species (increased levels of biodiversity) and have a different food webstructure. Organic agricultural ecosystems rely on the services of biodiversity for the recycling of nutrients through soils instead of relying on the supplementation of synthetic fertilizers. The model for ecological recycling agriculture adheres to the following principals:

- Protection of biodiversity.
- Use of renewable energy.
- Recycling of plant nutrients.

ECOSYSTEM ENGINEERS

The persistent legacy of environmental feedback that is left behind by or as an extension of the ecological actions of organisms is known as niche construction or ecosystem engineering. Many species leave an effect even after their death, such as coral skeletons or the extensive habitat modifications to a wetland by a beaver, whose components are recycled and re-used by descendants and other species living under a different selective regime through the feedback and agency of these legacy effects. Ecosystem engineers can influence nutrient cycling efficiency rates through their actions.

Earthworms, for example, passively and mechanically alter the nature of soil environments. Bodies of dead worms passively contribute mineral nutrients to the soil. The worms also mechanically modify the physical structure of the soil as they crawl about (bioturbation), digest on the moulds of organic matter they pull from the soil litter. These activities transport nutrients into the mineral layers of soil. Worms discard wastes that create worm castings containing undigested materials where bacteria and other decomposers gain access to the nutrients. The earthworm is employed in this process and the production of the ecosystem depends on their capability to create feedback loops in the recycling process.

Shellfish are also ecosystem engineers because they: 1) Filter suspended particles from the water column; 2) Remove excess nutrients from coastal bays through denitrification; 3) Serve as natural coastal buffers, absorbing wave energy and reducing erosion from boat wakes, sea level rise and storms; 4) Provide nursery habitat for fish that are valuable to coastal economies.

HISTORY

Nutrient cycling has a historical foothold in the writings of Charles Darwin in reference to the decomposition actions of earthworms. Darwin wrote about "the continued movement of the particles of earth". Even earlier, in 1749 Carl Linnaeus wrote in "the economy of nature we understand the all-wise disposition of the creator in relation to natural things, by which they are tted to produce general ends, and reciprocal uses" in reference to the balance of nature in his book *Oeconomia Naturae*. In this book he captured the notion of ecological recycling: "The 'reciprocal uses' are the key to the whole idea, for 'the death, and destruction of one thing should always be subservient to the restitution of another;' thus mould spurs the decay of dead plants to nourish the soil, and the earth then 'oers again to plants from its bosom, what it has received from them.'" The basic idea of a balance of nature, however, can be traced back to the Greeks: Democritus, Epicurus, and their Roman disciple Lucretius.

Following the Greeks, the idea of a hydrological cycle (water is considered a nutrient) was validated and quantified by Halley in 1687. Dumas and Boussingault (1844) provided a key paper that is recognized by some to be the true beginning of biogeochemistry, where they talked about the cycle of organic life in great detail. From 1836 to 1876,Jean Baptiste Boussingault demonstrated the nutritional necessity of minerals and nitrogen for plant growth and development. Prior to this time influential chemists discounted the importance of mineral nutrients in soil. Ferdinand Cohn is another influential figure. "In 1872, Cohn described the 'cycle of life' as the "entire arrangement of nature" in which the dissolution of dead organic bodies provided the materials necessary for new life. The amount of material that could be molded into living beings was limited, he reasoned, so there must exist an "eternal circulation" (ewigem kreislauf) that constantly converts the same particle of matter from dead bodies into living bodies."These ideas were synthesized in the Master's research of Sergei Vinogradskii from 1881-1883.

Variations in terminology

In 1926 Vernadsky coined the term biogeochemistry as a sub-discipline of geochemistry. However, the term nutrient cycle pre-dates biogeochemistry in a pamphlet on silviculture in 1899: "These demands by no means pass over the fact that at places where sufficient quantities of humus are available and where, in case of continuous decomposition of litter, a stable, nutrient humus is present, considerable quantities of nutrients are also available from the biogenic *nutrient cycle* for the standing timber. In 1898 there is a reference to the nitrogen cycle in relation to nitrogen fixing microorganisms. Other uses and variations on the terminology relating to the process of nutrient cycling appear throughout history:

- The term mineral cycle appears early in a 1935 in reference to the importance of minerals in plant physiology: "...ash is probably either built up into its permanent structure, or deposited in some way as waste in the cells, and so may not be free to re-enter the *mineral cycle*."
- The term nutrient recycling appears in a 1964 paper on the food ecology of the wood stork: "While the periodic drying up and reflooding of the marshes creates special survival problems for organisms in the community, the fluctuating water levels favor rapid *nutrient recycling* and subsequent high rates of primary and secondary production"
- The term natural cycling appears in a 1968 paper on the transportation of leaf litter and its chemical elements for consideration in fisheries management: "Fluvial transport of tree litter from drainage basins is a factor in *natural cycling* of chemical elements and in degradation of the land."
- The term ecological recycling appears in a 1968 publication on future applications of ecology for the creation of different modules designed for living in extreme environments, such as space or under sea: "For our basic requirement of recycling vital resources, the oceans provide much more frequent *ecological recycling* than the land area. Fish and other organic populations have higher growth rates, vegetation has less capricious weather problems for sea harvesting"

- The term bio-recycling appears in a 1976 paper on the recycling of organic carbon in oceans: "Following the actualistic assumption, then, that biological activity is responsible for the source of dissolved organic material in the oceans, but is not important for its activities after death of the organisms and subsequent chemical changes which prevent its *bio-recycling*, we can see no major difference in the behavior of dissolved organic matter between the prebiotic and post-biotic oceans."

Water is also a nutrient. In this context, some authors also refer to precipitation recycling, which "is the contribution of evaporation within a region to precipitation in that same region." These variations on the theme of nutrient cycling continue to be used and all refer to processes that are part of the global biogeochemical cycles. However, authors tend to refer to natural, organic, ecological, or bio-recycling in reference to the work of nature, such as it is used in organic farming or ecological agricultural systems.

RECYCLING IN NOVEL ECOSYSTEMS

An endless stream of technological waste accumulates in different spatial configurations across the planet and turns into a predator in our soils, our streams, and our oceans. This idea was similarly expressed in 1954 by ecologist Paul Sears: "We do not know whether to cherish the forest as a source of essential raw materials and other benefits or to remove it for the space it occupies. We expect a river to serve as both vein and artery carrying away waste but bringing usable material in the same channel. Nature long ago discarded the nonsense of carrying poisonous wastes and nutrients in the same vessels." Ecologists use population ecology to model contaminants as competitors or predators. Rachel Carson was an ecological pioneer in this area as her book *Silent Spring* inspired research into biomagification and brought to the worlds attention the unseen pollutants moving into the food chains of the planet.

In contrast to the planets natural ecosystems, technology (or technoecosystems) is not reducing its impact on planetary resources. Only 7% of total plastic waste (adding up to millions upon millions of tons) is being recycled by industrial systems; the 93% that never makes it into the industrial recycling stream is presumably *absorbed* by natural recycling systems In contrast and over extensive lengths of time (billions of years) ecosystems have maintained a consistent balance with production roughly equalingrespiratory consumption rates. The balanced recycling efficiency of nature means that production of decaying waste material has exceeded rates of recyclable consumption into food chains equal to the global stocks of fossilized fuels that escaped the chain of decomposition.

Pesticides soon spread through everything in the ecosphere-both human technosphere and nonhuman biosphere-returning from the 'out there' of natural environments back into plant, animal, and human bodies situated at the 'in here' of artificial environments with unintended, unanticipated, and unwanted effects. By using zoological, toxicological, epidemiological, and ecological insights, Carson generated a new sense of how 'the environment' might be seen.

Microplastics and nanosilver materials flowing and cycling through ecosystems from pollution and discarded technology are among a growing list of emerging ecological concerns. For example, unique assemblages of marine microbes have been found to digest plastic accumulating in the worlds oceans. Discarded technology is absorbed into soils and creates a new class of soils called technosols. Human wastes in the Anthropocene are creating new systems of ecological recycling, novel ecosystems that have to contend with the mercury cycle and other synthetic materials that are streaming into thebiodegradation chain. Microorganisms have a significant role in the removal of synthetic organic compounds from the environment empowered by recycling mechanisms that have complex biodegradation pathways. The effect of synthetic materials, such as nanoparticles and microplastics, on ecological recycling systems is listed as one of the major concerns for ecosystem in this century.

Technological recycling

Recycling in human industrial systems (or technoecosystems) differs from ecological recycling in scale, complexity, and organization. Industrial recycling systems do not focus on the employment of ecological food webs to recycle waste back into different kinds of marketable goods, but primarily employ people and technodiversity instead. Some researchers have questioned the premise behind these and other kinds of technological solutions under the banner of 'eco-efficiency' are limited in their capability, harmful to ecological processes, and dangerous in their hyped capabilities. Many technoecosystems are competitive and parasitic toward natural ecosystems. Food web or biologically based "recycling includes metabolic recycling (nutrient recovery, storage, etc.) and ecosystem recycling (leaching and *in situ* organic matter mineralization, either in the water column, in the sediment surface, or within the sediment."

8

Organic Farming and Crop Response and Quality of Products

ORGANIC FARMINGORGANIC FARMING

Organic agricultural practices are based on a maximum harmonious relationship with nature aiming at the non-destruction of the environment. The developed nations of the world are concerned about the spreading contamination of poisonous chemicals in food, feed, fodder and fibre. Naturally, organic farming system is looked upon as one of the means to remedy these maladies there. However, the major problem in India is the poor productivity of our soils because of the low level content of the organic matter.

The efficiency of the organic inputs in the promotion of productivity depends on the organic contents of the soil. There were many resemblances of organic farming principles in the traditional agriculture of India. But the former gives a more open and verifiable scientific foundation than the latter.

HEALTHY FOODS

A study conducted in USA on the nutritional values of both organic and conventional foods found that consumption of the former is healthier. Apples, pears, potatoes, corn, wheat and baby foods were analysed to find out 'bad' elements such as aluminum, cadimum, lead and mercury and also 'good' elements like boron, calcium, iron, magnesium sellenium and zinc. The organic food, in general, had more than 20 per cent less of the bad elements and about 100 per cent more of the good elements.

IMPROVEMENT IN SOIL QUALITY

Soil quality is the foundation on which organic farming is based. Efforts are directed to build and maintain the soil fertility through the farming practices. Multicropping, crop rotations, organic manures and pesticides, and minimum tillage are the methods employed for the purpose. Natural plant nutrients from green manures, farmyard manures, composts and plant residues build organic content in the soil. It is reported that soil under organic farming conditions had lower bulk density, higher water holding capacity, higher microbial biomass carbon and nitrogen and higher soil respiration activities compared to the conventional farms. This indicates that sufficiently higher amounts of nutrients are made available to the crops due to enhanced microbial activity under organic farming. The effect of organic cultivation on soil fertility as reported at the farm of Central Institute for Cotton Research, Nagpur is given in Table.

Table. Organic carbon and available P contents

Year	Organic carbon (%)	Organic Available P (kg/ha)	Organic carbon (%)	Non-organic Available P (kg/ha)
June 1993	0.38	12.1	0.38	12.1
Feb 1994	0.40	12.6	0.36	12.0
Feb 1995	0.46	14.5	0.35	12.9
Feb 1996	0.52	15.0	0.38	12.0

INCREASED CROP PRODUCTIVITY AND INCOME

Field trials of organic cotton at Nagpur revealed that during the conversion period, cotton yield was low compared to the conventional and integrated crop management. However, the yields of organic cotton started rising from third year.

Cotton yields under organic, conventional and the mixed systems were 898, 623 and 710 kg/ha respectively at the end of the fourth year of the cultivation. The yield of soyabean under organic farming was also the highest compared to the other two systems.

The Central Institute for Cotton Research, Nagpur conducted a study of economics of cotton cultivation in Yavatmal district of Maharashta. The cost of cultivation of cotton was lower in the organic farming than in the modern system. The low costs were due to the non-use of fertilizers and chemical insecticides.

As a result of the low yields during the conversion period, the net income from the organic farm was lesser than the conventional farm. But the yield under organic method increased progressively equalling it to that of the conventional system by the sixth year.

The input costs were low under organic farming and with a 20 per cent of premium prices of output, the net income increased progressively from fourth year under organic fanning. The appreciation of net income from organic cotton cultivation by the sixth year was 80 per cent over the conventional crop.

Results reported from 1050 field demonstration cum trials under the National Project on Development and Use of Biofertilizers in different parts of the country show* an increase of 4 per cent in yield in plantation crops, 7 per cent in fruit crops, 9 per cent in wheat and sugarcane, 10 per cent in millet and vegetable, 11 per cent in fibre, condiments and spice crops, 14 per cent in oilseeds and flowers and 15 per cent in tobacco.

A study of 100 farmers in Himachal Pradesh during a period of 3 years found that the total cost of production of maize and wheat was lower under organic farming and the net income was 2 to 3 times higher. Both productivity and premium prices contributed to the increased profitability. Another study of 100 farmers of organic and conventional methods in five districts of Karnataka indicated that the cost of organic farming was lower by 80 per cent than that of the conventional one.

Table. The Cost Benefit Ratios Described for Various Crops

Crops	Organic	Inorganic
Groundnut	1:1.26	1:1.31
Jowar	1:1.36	1:1.28
Cotton	1:1.34	1:1.24
Coconut	1:1.70	1:1.31
Banana	1:3.66	1:2.82

Low Incidence of Pests

The study of the effectiveness of organic cotton cultivation on pests at the farm of Central Institute for Cotton Research, Nagpur revealed that the mean monthly counts of eggs, larva and adults of American BoUworm were far lesser under organic farming than under the conventional method.

Bio-control methods like the neem based pesticides to Ti-ichoderma are available in the country. Indigenous technological products such as Panchagavya which was experimented at the University of Agricultural Sciences, Bangalore found to control effectively wilt disease in tomato.

Employment Opportunities

Organic farming requires more labour input than the conventional farming system. Thus, India which has a very large amount of labour unemployment and under employment will find organic farming an attraction.

Moreover, the problem of periodical unemployment will also get mitigated because of the diversification of the crops with their different planting and harvesting schedules resulting in the requirement of a relatively high labour input.

Indirect Benefits

Several indirect benefits from organic farming are available to both the farmers and consumers. While the consumers get healthy foods with better palatability and taste and nutritive values, the farmers are indirectly benefited from healthy soils and farm production environment. Eco-tourism is increasingly becoming popular and organic farms have turned into such favourite spots in countries like Italy. Protection of the ecosystem, flora, fauna and increased biodiversity and the resulting benefits to all human and living things are great advantages of organic farming which are yet to be properly accounted for.

PROPOSED OBJECTIVES

The broad objectives of organic farming in India can be the following in the light of the discussions on the adverse effects of the conventional farming system the country practiced for about 30-40 years and the potential benefits of the organic methods:

- Sustainable agriculture
- Increasing agriculture production
- Food self-sufficiency
- Environmental protection
- Conservation of natural resources
- Rural development.

ORGANIC FARMING PRINCIPLES AND PRACTICES

The roots of organic agriculture developed from differing systems of thought, philosophies of life and agro-political motivations. One thing they all have in common is the desire to form a method of production capable of generating healthful foodstuffs, while limiting any damaging effects on the natural ecosystem. It has in the meantime been scientifically proven beyond doubt that organic farming systems are the most environmentally-friendly, and thus sustainable, agricultural methods. This method of production actively assists in preserving eco-systems and the variety of species, protecting the soil, keeping the water clean and reducing the impact of agriculture on the atmosphere. Organic agriculture is concerned not only with leaving out technical production aids, such as pesticides or synthetically-produced chemical, mineral fertilizers, or simply replacing them with aids permitted in an organic farming system.

But is rather more a holistic cultivation system whereby an agricultural site is viewed as an organism. This method of planting has little in common with the "Ancient's agricultural system", but has been developed from a process based on technical-biological progress. Organic agriculture consciously avoids trying to maximise the yield per cultivation area. The total productivity of a farm (including the ecological aspects), optimally adapted to the site conditions, is the most important aspec.

The following basic principles should be closely followed:

- Sustaining and improvement of the soil.

- Realisation (near as possible) of nutrient re-cycling (farm, village, region). Intensive use of legumes/leguminous trees to provide nitrogen supply.
- Biological plant protection through prevention.
- Biodiversity of crop varieties and species grown.
- Site and species appropriate animal husbandry.
- Prohibition of Genetic Engineering and products thereof.
- Maintenance of the surrounding natural landscape (sustainable eco-agro systems).
- Least possible consumption of non-renewable energy and resources.
- Ban on synthetic, chemical fertilizers, plant protection, storage and ripening means as well as hormones and synthetic growth regulators (also harmful processing aids in food processing).

Each individual farmer, whether in the tropics or more temperate zones, must decide for himself exactly how he can practically apply these basic principles to his daily work. A variety of solutions can be developed, depending upon the specific site and farm conditions. Intensive levels of specialisation (monocultures) within a farm (a village or region) should be avoided to the same degree as the destruction of an intact natural eco-system through agriculture (e.g., slash and burn method in tropical rainforests).

Rather, sustainable eco-agro systems should be aspired to, which are integrated with the flora and fauna present at the site. Finally, it should also be noted that a consensus exists within the International Federation of Organic Agriculture Movements (IFOAM), that organic agriculture must also take the socio-economical conditions of the people in a region, village or on a farm into consideration. The degree to which organic agricultural systems can thrive in a region depends to a large extent upon the opportunities the people who live there have to participate in them.

In this respect, due to its inherent diversity, organic agriculture can be applied as an instrument with a range of uses in rural development strategies. By utilising this instrument, developmental perspectives can be generated for a rural population.

HISTORICAL PERSPECTIVE

Understanding why a process or practice is done a certain way, or why we think the way we do can often be accomplished by looking at the historical evolution of human learning and perception. The history of agriculture and the human perception of how plants grow are both well documented and can help us be better farmers.

Let's start with observing the growth of a seed. We plant a little seed and in a short time we have a large plant. The seed weighs a fraction of an ounce and in a short time the plant weights several pounds. What makes up this weight and were did it come from? Based on what we can see with our eyes and conclude based on observation, what is necessary for a plant to grow?

Based on observing a seed germinate and grow in a garden, we can say the plant needed soil, water, light, and a reasonable temperature. A seed on poor ground will not grow very well. A seed without water will not grow. A seed without light will only grow a short time. And the seed will not grow for very long if it is too cold or too warm.

Where did the mass or weight of the plant come from? We know that some of the plant is made up of water. We know that the plant required light, but we don't have any evidence just looking with the human eye that light energy was transformed into solid matter. Soil and nutrients were required, but did the weight come from the soil? We could observe that a seed growing in soil that has decaying plant material or animal manure will usually grow larger and faster than one growing in soil without organic matter added. It is not that long ago, about 150 years, that the main theory of plant growth was the "humus" theory. Those observing plants assumed that the weight of the plant came from the soil. The idea was that the root somehow ingested the soil or something in the soil, perhaps "humus" and the plant was assembled from these building blocks. Plants were thought to be like animals, only eating from the soil.

Only this time, let's take a container of soil and put the seed in it. We will add water and keep track of how much we add. After the plant grows, we can weigh the plant. If the weight comes from the soil, we will see a decrease in the weight of the soil at the end of the experiment. This has been done before, over 400 years ago for likely the first time that we know about.

The answer was that the plant grew but the soil stayed about the same, only a few ounces of soil were lost, and that could have been an error. So where did the plant come from? It must have been the water?

Let's try growing the seed, or a seedling in just water. A seedling does not grow very well at all in purified water, but if some garden soil is added to the water, the seedling grows. Perhaps there is something from the soil? The thinking continued that the soil was a prime supplier of materials for plant growth.

What happens if we grow the plant in a giant plastic bag or air tight glass case or greenhouse? After a while the plant doesn't grow very well if at all. What if we pump some fresh air into the bag or case? Suddenly the plant starts growing better. Is there something in the air that the plant needs? We know the answer is yes, because now we know the air is made up of not one gas but several gasses, including nitrogen (~78 per cent), oxygen (~20 per cent), and carbon dioxide (~0.034 per cent). The identification of oxygen and the purification ability of plants was identified around 1800 (Priestly and others.).

An animal in a closed, airtight container dies and a plant in an airtight container may survive but will not grow. Put them together and they both can thrive. The animal gives off something that would make the plant grow and the plant gives off something the animal needs. Carbon dioxide from the animal to the plant (respiration) and oxygen from the plant to the animal (photosynthesis). Where does the carbon in the food we eat go? Some is incorporated into our body. Some leaves the body in urine or feces. Is there any other way carbon leaves the body? Yes in our breath. Instead of the approximately 350 parts of carbon in fresh air, our breath has approximately 20,000 parts or 80 times more. The exhaling of carbon dioxide is common for animals right down to single cell bacteria. Let's try a different approach. Let's take the plant or plants and dry it and get rid of all the water. Instead of 100 pounds of fresh plant, we now have 10 pounds dry matter. We evaporated about ninety pounds or 12 gallons of water. There is still 10 pounds of plant. We could take this dry plant and burn it. The plant disappears or goes up in smoke, except for about 1 pound of ashes. Where did the other 9 pounds go? What was in that smoke and gas that formed as the plant burned? What do we get when we burn wood or organic material? Heat, carbon dioxide and water, made up of carbon, hydrogen and oxygen.

What was in the ashes? All the other elements the plant needed to grow and extracted from the soil. That little bit of soil that was missing was not an error in weighing, we have now recovered most of it in the form of ashes. In the ashes would be potassium, phosphorus, calcium, magnesium, sulphur, iron, manganese, zinc, copper, boron, molybdenum, chloride, silicon, nickel and a small amount of other elements. (Ashes once seen as bones of plants.) The nitrogen would be lost to the atmosphere.

Looking at what elements and minerals were in plants, together with the evolving science of chemistry, led German scientists, particularly one organic chemist by the name of von Leibig, to state that the plant took some things from the soil, but most of the carbon came from the air, not the soil. It was around this time that the assumption developed that only the minerals extracted from the soil had to be replaced in order to maintain yields. Artificial manures made from salts were developed and even patented.

Later, around 1860, it was demonstrated that plants could be grown in water if the correct salts were added to the water. This is the process that later became known as hydroponics. *Much of our current approach based on chemical fertilizers used in conventional agriculture is based on these observations and this way of thinking.*

Other than naturally occurring in the soil, fertilizer elements can either be mined from the earth or manufactured. Most of the N used in modern industrial agriculture is taken from the atmosphere. Under high temperatures and pressures, obtained using fossil fuels such as natural gas, the nitrogen molecule (N2) can be combined with hydrogen (H2) to form an ammonium molecule NH4, which is useful to plants. It can also be mixed with oxygen and chemically converted to nitrate (NO3), which can be used by plants. How did we learn to take nitrogen from the atmosphere? Mostly from research to get nitrogen to make explosives. Large scale nitrogen production traditionally has been looked at as a valuable outcome of World War I.

After the war, the munition plants and the people working there needed something to do and someone figured out they could make fertilizers. This lead to the eventual development and application of many chemicals now used in conventional agriculture (1938 USDA yearbook of Agriculture). The rapid growth in agriculture that occurred with the post war (WWII) baby boom is in a large part the result of increased nitrogen fertilizer and how it was used to increase yield and lower food costs.

So, if most of the plant comes from the atmosphere and some salts from the soil, if the plant does not extract organic matter from the soil for growth, what is the importance of having organic matter in the soil? After all, we can grow plants in water and fertilizer solution using hydroponics. Even von Leibig was aware and fascinated by the problems of Virginia tobacco farmers in the new world. It seems after 100 years of growing wheat and tobacco without any addition of manure, the crops were not growing and the land was being abandoned. It is important to recognise that von Leibig was a chemist, and had little understanding of farming. It has also been reported that late in his career he realised the mistake he had made in making it sound like organic matter was not important, but it was too late to undo the damage.

CONSERVATION TILLAGE AND ORGANIC FARMING

Organic agriculture is often characterized as addicted to maximum tillage — with growers using every opportunity to lay the land bare with shovel, plow, or rototiller. This image has been magnified through the popularity of small-scale organic systems like the French Intensive and Biointensive Mini Farming models that espouse double- and triple-digging to create deep rooting beds. While appropriate to such intensive systems, this degree of cultivation is not characteristic of organic agriculture in general. It may surprise some to learn that a large number of organic producers are not only interested in conservation tillage, they have adopted it. This will be a surprise because many believe that conservation tillage always requires herbicides.

The interest in conservation tillage among organic producers in the Cornbelt was well documented in the mid-1970s by Washington University researchers. They noted that the vast majority of organic farmers participating in their studies had abandoned the moldboard plow for chisel plows. Plowing with a chisel implement is a form of *mulch tillage*, in which residues are mixed in the upper layers of the soil, and a significant percentage remains on the soil surface to reduce erosion. Furthermore, a notable number of organic farmers had gone further to adopt ridge-tillage, a system with even greater potential to reduce erosion. It was especially interesting to note that the use of these conservation technologies was almost nil among neighboring conventional farms at the time. Organic growers were actually pioneers of conservation tillage in their communities. Among the more well-known of these pioneers were Dick and Sharon Thompson of Boone, Iowa. Their experiences with ridge-tillage and sustainable agriculture became the focus of a series of publications titled *Nature's Ag School*. These were published by the Regenerative Agriculture Association — the forerunner to the Rodale Institute. They are now, unfortunately, out of print.

Research continues to open up new possibilities in conservation tillage for organic farms. New strategies for mechanically killing winter cover crops and planting or transplanting into the residue without tillage are being explored by a number of USDA, land-grant, and farmer researchers. Notable among these is the work being done by Abdul-Baki and Teasdale at the USDA in Beltsville, Maryland — transplanting tomato and broccoli crops into mechanically killed hairy vetch and forage soybeans. There are also the well-publicized efforts of Pennsylvania farmer Steve Groff, whose no-till system centres on the use of a rolling stalk chopper to kill cover crops prior to planting. Systems like Groff's and Abdul-Baki's are of particular interest because close to 100% of crop residue remains on the soil surface – providing all the soil conservation and cultural benefits of a thick organic mulch.

After blind cultivation, subsequent weed control operations in larger-scale systems can make use of advances in tillage equipment such as rolling cultivators, finger weeders, and torsion weeders that allow tilling close to the plant row. Smaller-scale operations often use wheel hoes, stirrup hoes, and other less capital-intensive hardware.

Determining the amount, the timing, and the kind of tillage to be done can be a balancing act for the organic grower, but experience and observation over time lead to proficiency. There are downsides to tillage, however, and most organic growers are well aware of them. The most obvious of these is the dollar cost; organic farmers are as concerned as their conventional counterparts about costs of production and strive to minimize expensive field operations. There is also a cost to the soil and environment. Every tillage operation aerates the soil and speeds the decomposition of the organic fraction. While this may provide a boost to the current crop, it can be overdone and "burn up" the humus reserves in the soil. Excessive tillage can also be directly destructive to earthworms and their tunneling, reducing their benefits to the land. There is also the danger of compaction, even when field operations are well timed.

PRACTICE OF MULCHING

Mulching is a practice often used by organic growers. Traditionally, it entails the spreading of large amounts of organic materials — straw, old hay, wood chips, etc. — over otherwise bare soil between and among crop plants. Organic mulches regulate soil moisture and temperature, suppress weeds, and provide organic matter to the soil. Mulching is most appropriate to small, intensive operations with high-value annual or fruit crops. A few systems of no-till organic gardening have evolved from the concept of deep, *permanent* mulching. Among these are the well-known Ruth Stout method and *Synergistic Agriculture* — a raised bed system developed by Emilia Hazelip, who adapted concepts from *Permaculture* and the ideas of Masanobu Fukuoka. Mark Cain and Michael Crane, co-owners of Dripping Springs Gardens — an intensive market gardening operation in Arkansas — have adapted Emilia's system to their farm with considerable success.

Plastic mulch, as long as it is removed at end of growing or harvest season, is also permitted in certified organic production. Its use allows larger acreage to be brought more easily under herbicide-free management, though there are serious issues to be addressed.

High-Input Organic Agriculture

At the beginning of this publication, organic farming was described as a system that uses a minimum of off-farm inputs. While that describes most of organic agriculture as it is currently practiced in the U.S., certified organic farming can also entail much greater reliance on off-farm inputs. Intensive annual strawberry and vegetable systems under plasticulture are good examples. In these systems, traditional rotations and soil building practices are usually employed, followed by clean cultivation and the laying of plastic mulch and drip irrigation tape on shaped beds. During the season, large amounts of soluble organic fertilizers — typically fish-based — are fed to the crop through the drip system (*i.e.*, organic fertigation). At the end of the season, all plastics must be removed from the field, and it is returned to more standard organic management.

Ideally, an off-season cover crop will be planted. Such systems are often exceptionally productive and economically attractive, when organic premiums are good. The high cost of soluble organic fertilizer (typically hundreds of dollars/acre), however, plus the marginally higher cost of pest controls, make such systems largely non-competitive in the conventional marketplace. The labeling of such high-input systems as organic presents a paradox for many proponents of organic agriculture.

It is unclear whether these technological advancements reflect the kind of farming most practitioners and supporters of organics think of as truly "organic." To begin with, the research citing environmental and economic benefits has largely been done on low-input organic systems; it is questionable whether similar findings would be made about high-input systems, especially regarding environmental matters. Of particular note, while low-input organic systems are documented as being more resistant to erosion, fields under plastic mulch are reported to be fifteen times more erodible. Traditional organic farms leach minimal amounts of nitrogen into tile or groundwater; the losses from fields loaded with high levels of soluble organic fertilizers is certain to be greater, but how much greater is unknown. The fossil fuel energy involved in plastic manufacture, transportation, and application may or may not be compensated by reductions in tractor fuel use.

Finally, the lowered capital investment required to produce a crop by traditional organic methods makes this form of farming more accessible to resource-poor farmers and entails less risk in years of crop failure or lack of premiums. These factors are less certain in a high-input system. A further consideration is the issue of plastic disposal following removal. At this time, there are few to no options for recycling, and landfills are the fate of plastic mulches at the end of each season.

While it is unwise to rush to judgement regarding high-input organic farming, it is clear that some adaptations will need to be made, if the traditional character and sustainability benefits of organic farming are to be preserved.

Fire

While fire can be used in a number of ways in organic agriculture, the area of greatest interest is *flame* or *thermal weeding*. In its most common application, torches mounted on a tractor toolbar throw a hot flame at the base of mature (*i.e.*, heat-resistant) plants, over the inter-row area, or both. Tractor speed is adjusted so that weeds are not *burned* so much as *seared*. Searing is sufficient to kill most seedling weeds and uses less fuel. Liquid propane (LP) gas is the fuel most commonly used, though alternatives such as alcohol and methane offer the possibility of on-farm sources.

Supplemental Fertilization

In many organic systems, crop rotation, manuring, green manuring, along with enhanced biological activity in the soil, provide an abundant supply of plant-essential minerals annually. This is especially true on naturally deep and rich prairie soils. It is less true on poorer soils and on those that have been heavily exploited through non-sustainable farming practices. To correct mineral deficiencies in organically managed soils, organic growers often apply ground or powdered rock minerals.

The most commonly used rock mineral is high-calcium aglime. Dolomitic limestone, various rock phosphates, gypsum, sulfate of potash-magnesia, and mined potassium sulfate are also common. These are all significant sources of primary (P, K) and/or secondary (Ca, Mg, S) plant nutrients. The savvy organic grower applies significant amounts of these materials only with the guidance of regular soil testing.

Less common are other rock powders and fines that are limited sources for the major nutrients but are rich in micronutrients or have some other soil-improving characteristic. Among these are glauconite (greensand), glacial gravel dust, lava sand, Azomite®, granite meal, and others.

Supplementary nutrients that include nitrogen are often provided in the form of animal or plant products and by-products such as fish emulsion, blood meal, feather meal, bone meal, alfalfa meal, and soybean meal. Most of these products also supply some organic matter, though that is not the primary reason they are used.

Evaluating Tools and Practices

One basis on which to evaluate the tools or practices one chooses for an organic operation is whether or not they contribute to biodiversity — biodiversity being one of the principal characteristics of a sustainable organic agriculture.

Crop rotation, cover cropping, farmscaping, companion planting, and intercropping are outstanding examples of practices that contribute to biodiversity. They therefore contribute to the long-term stability and sustainability of the farm agroecosystem. Composting and manuring likewise contribute to biodiversity, but since the diversity they promote is mostly in soil biota, it is rather less obvious to the casual observer.

On the other end of the spectrum are practices such as tillage, cultivation, thermal weeding, solarization, and plastic mulching. Such tools significantly reduce diversity in the field and tend to move the system in a less sustainable direction. This, however, does not necessarily make these practices bad choices.

Organic farming has often been called *natural* farming, as it tries to mimic the processes of nature in producing crops and livestock. However, the analogy goes only so far, since most agricultural systems are characterized by a struggle between human and nature, each with a clear notion of what plants and animals the land *ought* to support and in what proportions. The farmer is typically in the position of "holding back" the natural succession of plant and animal species through the use of diversity-reducing tools and practices. The ideal is to bring about an agricultural system in which the long-term direction emphasizes diversity and sustainability. Among the best visualizations of this ideal are those emerging from the Permaculture movement.

Balanced Nutrition

Whether or not it has been customary in the past, organic growers are encouraged to have periodic soil testing done on their fields. How the results of a soil test are used, however, can vary considerably among farmers, depending on their personal philosophy and management skill. While there is certainly a segment of the organic farming community that has no faith in soil audits, most growers use them as a means to monitor progress in building their soils, to identify nutrient deficits, and to guide supplementary fertilization. While there are many ideas about fertilization guidelines, there are two schools of thought that dominate.

The first is commonly known as the *sufficiency* approach or model. Though somewhat oversimplified, the following are among its significant characteristics:

- Annual fertilizer additions of P and K are based in good part on how much the crop is expected to remove from the soil at harvest
- Additional amounts of P and K are recommended based on keeping the soil nutrient reserves at a particular level
- Lime is added to the soil based on pH
- Little to no attention is paid to nutrient balance or to the levels of secondary nutrients Ca and Mg.

The second approach is referred to variously as *cation nutrient balancing*, the *Albrecht system*, the *CEC*, or the *base saturation* approach. It differs from the sufficiency approach in that fertilizer and lime recommendations are made based on an idealized ratio of nutrients in the soil and its capacity to hold those nutrients against leaching.

Cation nutrient balancing is more popular among practitioners of organic farming and sustainable agriculture in general than it is among conventional growers. However, there is no universal agreement on which approach is most appropriate for organic management.

A wide range of other products —humates, humic acids, enzymes, catalyst waters, bioactivators, surfactants, to name a few — are also acceptable in organic crop production. However, the tradeoff between out-of-pocket cost and efficacy of such materials is often challenged by conventional and organic growers alike. Organic growers are encouraged to experiment, but to do so in a manner that allows the actual results to be measured.

Biorational Pesticides

While, in principle, any pesticide use is discouraged in organic systems, a rather wide range of biorational pesticides is permitted. The frequency of pesticide use varies considerably with crop and location. For example, there is virtually no use of pesticides on organic row crops in the Cornbelt. By contrast, organic tree fruits in the Midsouth routinely receive heavy applications of several fungicides and insecticides allowable in organic production.

The pesticides permitted in organic farming fall predominantly into several classes.

- *Minerals:* These include sulfur, copper, diatomaceous earth, and clay-based materials like Surround®.
- *Botanicals:* Botanicals include common commercial materials such as rotenone, neem, and pyrethrum. Less common botanicals include quassia, equisetum, and ryania. Tobacco products like Black-Leaf 40® and strichnine are also botanicals but are prohibited in organic production due to their high toxicity.
- *Soaps:* A number of commercial soap-based products are effective as insecticides, herbicides, fungicides, and algicides. *Detergent*-based products are *not* allowed for crop use in organic production.
- *Pheromones:* Pheromones can be used as a means to confuse and disrupt pests during their mating cycles, or to draw them into traps.
- *Biologicals:* One of the fastest-growing areas in pesticide development, biopesticides present some of the greatest hope for organic control of highly destructive pests. Among the most well-known biopesticides are the *Bacillus thuringiensis* (Bt) formulations for control of lepidopterous pests and Colorado potato beetle.

NEED FOR ORGANIC FARMING

The need for organic farming in India arises from the unsustaina-bility of agriculture production and the damage caused to ecology through the conventional farming practices. The present system of agriculture which we call 'conventional' and practiced the world over evolved in the western nations as a product of their socio-economic environment which promoted an over riding quest for accumulation of wealth. This method of farming adopted by other countries is inherently self destructive and unsustainable.

The modern farming is highly perfected by the Americans who dispossessed the natives of their farms right from the early period of the new settlers in US. The large farms appropriated by the immigrants required machines to do the large scale cultural operations. These machines needed large amount of fossil fuels besides forcing the farmers to raise the same crops again and again, in order to utilize these machines to their optimum capacities. The result was the reduction of bio-diversity and labour. The high cost of the machines necessitated high profits, which in turn put pressure to raise productivity. Then, only those crops with high productivity were cultivated which needed increased quantities of fertilizers and pesticides. Increasing use of pesticides resulted in the damage to environment and increased resistance of insects to them. Pesticides harmed useful organisms in the soil. The monoculture of high yielding seeds required external inputs of chemical fertilizers. The fertilizers also destroy soil organisms. They damage the rhizobia that fix nitrogen and other micro organisms that make phosphates available to plants. The long term effect was reduction of crop yields. The damaged soil was easily eroded by wind and water. The eroding soil needed use of continuously increasing quantities of fertilizers, much of which was washed/leached into surface and underground water sources.

The theme of consumer welfare has become central in the economic activities in the developed countries in the world. Sustainable agriculture based on technologies that combine increased production with improved environmental protection has been accepted as absolutely essential for the maximization of the consumer welfare.

The consumers are increasingly concerned about the quality of the products they consume and food safety has become a crucial requirement. Safety, quality and hygienic standards are increasingly being made strict. The mad cow disease and the question of genetically modified food production are the recent instances, which made the countries to tighten the laws. Mycotoxln contamination, unacceptable levels of pesticide residues and environment degradation are the problems on which the attention is centred. Keeping the interests of the consumers, the European Union has taken tough measures including criminal prosecution to ensure food safety. Another area to increase the consumer welfare is promotion of the eco-friendly methods in agriculture. No-till, or conservation agriculture, lower input approaches of integrated pest or nutrient management and organic farming are some of them.

The Indian agriculture switched over to the conventional system of production on the advent of the green revolution in the 1970s. The change was in the national interest which suffered set backs because of the country's over dependence on the foreign food sources. The national determination was so intense that all the attention was focused on the increase in agriculture production.

The agriculture and allied sectors in India provide employment to 65 per cent of the workers and accounts for 30 per cent of the national income. The growth of population and the increase in income will lead to a rise in demand for foodgrains as also for the agricultural raw materials for industry in the future. The area under cultivation, obviously, cannot be increased and the present 140 million hectares will have to meet the future increases in such demands. There is a strong reason for even a decline in the cultivated area because of the urbanization and industrialization, which in turn will exert much pressure on the existing, cropped area. Science and technology have helped man to increase agricultural production from the natural resources like land. But the realization that this has been achieved at the cost of the nature and environment, which support the human life itself, is becoming clear.

It has been fully evident that the present pattern of economic development, which ignores the ecology and environment, cannot sustain the achievement of man without substantial erosion of the factors that support the life system of all living things on the Earth. The evidence of the ill effects of development is well documented. As said earlier, we in India have to be concerned much more than any other nation of the world as agriculture is the source of livelihood of more than 6-7 million of our people and it is the foundation of the economic development of the country.

There were times when people lived close to nature with access to flora and fauna in healthier and cleaner surroundings. One has to look back at our present metropolitan cities or other large towns before the past fifty years as recorded in history/memmories of the present elder generation to see the striking differences in the surroundings in which the people lived there. Land, water and air, the most fundamental resources supporting the human life, have degraded into such an extent that they now constitute a threat to the livelihood of millions of people in the countiy. Ecological and environmental effects have been highly publicised all over the world. Many times, these analysis have taken the shape of doomsday forecasts. Powerful interests in the developed western countries have also politicised these issues to take advantage of the poor nations of the world. Efforts to impose trade restrictions on the plea of environment protection are a direct result of these campaigns. But we have to recognize that the abysmal level to which we have degraded our resources,requires immediate remedial measures without terming the demand for them as the ploys of the rich nations to exploit the poor.

Another turn of the events has been the blame game for ecological problems stated at the Earth Summit and other international conferences. The developed countries, it is true, are to a great extent instrumental to degrade the environment. However, the poorer countries of the world including India cannot delay or ignore the need for remedial measures, which are to be effectively implemented. We cannot gloss over the fact that we have also contributed to the degradation of ecology; look at the droughts and floods, disappearance of forests, high noise level and air pollution in the cities which are our own creations.

Organically cultivated soils are relatively better attuned to withstand water stress and nutrient loss. Their potential to counter soil degradation is high and several experiments in arid areas reveal that organic farming may help to combat desertification. It is reported that about 70 hectares of desert in Egypt could be converted into fertile soil supporting livestock through organic and biodynamic practices. India, which has some areas of semi-arid and arid nature, can benefit from the experiment. The organic agriculture movement in India received inspiration and assistance from IFOAM which has about 600 organizational members from 120 countries. All India Federation of Organic Farming is a member of IFOAM and consists of a number of NGOs, farmers' organisations, promotional bodies and institutions.

The national productivity of many of the cereal crops, millets, oilseeds, pulses and horticultural crops continues to be one of the lowest in the world in spite of the green revolution. The fertilizer and pesticide consumption has increased manifold; but this trend has not been reflected in the crop productivity to that extent. The country's farming sector has started showing indications of reversing the rising productivity as against the increasing trend of input use.

The unsustainability of Indian agriculture is caused by the modern farming methods which have badly affected/damaged production resources and the environment.

Affects of Modern Farming Technology

The role of agriculture in economic development in an agrarian country like India is a pre-dominant one. Agriculture provides food for more than 1 billion people and yields raw materials for agro-based industries. Agricultural exports earn foreign exchange. Modernization of Indian agriculture began during the mid-sixties which resulted in the green revolution making the country a foodgrain surplus nation from a deficit one depending on food imports. Modern agriculture is based on the use of high yielding varieties of seeds, chemical fertilizers, irrigation water, pesticides, etc., and also on the adoption of multiple cropping systems with the extension of area under cultivation.

But it also put severe pressure on natural resources like, land and water. However, given the continuous growth of modern technology along with the intensive use of natural resources, many of them of non renewable, it is felt that agriculture cannot be sustainable in future because of the adverse changes being caused to the environment and the ecosystem. The environmental non-degradable nature of the agricultural development and its ecological balance have been studied in relation to the modem Indian farming system by experts which shows exploitation of land and water for agriculture, and the excessive use of chemicals.

CHEMICAL CONTAMINATIONCHEMICAL CONTAMINATION

Fertilizers

Consumption of chemical fertilizers {N,P,K) has been increasing in India during the past thirty years at a rate of almost half a million tonnes on an average, a year. It was only 13.13 kg/ha in 1970-71, 31.83 kg/ha in 1980-81 and 74.81 kg/ha in 1995-96. It shot up to about 96 kg/ha during 1999-2000. Table shows the consumption of fertilizers in India from 1970-71 to 2001-02.

Table. Consumption of chamical fertilizers in India.

Sr. No.	Year	Consumption (M. Tonnes)	Consumption (Kg/ha)
1	1970-71	2.18	13.13
2	1980-81	5.52	31.83
3	1990-91	12.54	67.49
4	1991-92	12.73	69.84
5	1992-93	12.15	65.53
6	1993-94	12.24	66.69
7	1994-95	13.56	73.12
8	1995-96	13.58	74.81
9	1996-97	14.31	76.70
10	1997-98	16.19	86.80
11	1998-99	16.80	89.80
12	1999-00	18.07	95.60
13	2000-01	16.71	NA
14	2001-02	17.54	NA

The present use of about 96 kg of fertilizers per ha in India appears to be modest compared to the advanced countries. Currently about 80 per cent of the fertilizer is consumed in only about 120 districts constituting less than 33 per cent of the gross cultivating area. Experts point out that the efficiency of fertilizer use in India is only 30-35 per cent as the balance 65-70 per cent reaches the under ground water. The intensity of their use in a few regions and a few crops are causes of serious concern to human health, soil, water, environment and thus to the sustainability of agriculture production in the country. It is true that the increasing use of fertilizer at high rates has boosted agricultural production in the country. But it has also caused adverse impact on soil and water as well as environment. Several studies on the effects of high level of fertilizer application on soil health have confirmed the adverse impacts. Both drinking and irrigation water wells in large numbers have been found contaminated with nitrates, some of them are having even 45 mg per litre, well above the safe level.

Long term continuous use of high doses of chemical fertilizers badly affects the physical, chemical and biological properties of the soil. A study at the University of Agricultural Sciences, Bangalore confirmed the deterioration of soil health because of the reduction in water holding capacity, soil pH, organic carbon content and the availability of trace elements such as zinc in case of ragi crop even with the application of normal doses of fertilizer in the long run. In the long run, increasing nitrogenous fertilizer use leads to the accumulation of nitrates in the soil.

Table. Content of heavy metals in fertilizers and sludges (Matal: mg/kg)

Source	Cd	Cr	Cu	Pb	Zn
Ammonium nitrate	1.1	2.5	3.6	5.4	11.7
Super phosphate	16.6	157.0	22.6	20.6	244.0
NPK fertilizers (8-10-18)	4.9	54.3	8.3	3.2	97.5
Sewage sludge	20.0	500.0	250.0	700.0	3000.0

The application of sulphatic fertilizers leaves sulphates in the soil. Rainfall and excessive use of irrigation water cause these chemicals to change the alkaline or acidic nature of the soil. The nitrates go to the rivers, wells, lakes etc. And also leak into the drainage system which goes into the drinking water contaminating the environment. It also causes depletion of the ozone layer adding to the global warming. Use of nitrogen in the form of ammonium sulphate in the rice crop emanates ammonia polluting the atmosphere. The heavy metals present in the fertilizers and sewage sludge leach into ground water. Table shows the content of some heavy metals in fertilizers and sludges.

Pesticides

The use of chemical pesticides began with the discovery of toxicological properties of DDT and HCH during the Second World War. Many chlorinated hydrocarbon insecticides like aldrin, dieldrin, toxaphane, chlordane, endosulfan, etc. came into the market during the second half of the last century. Simultaneously, organophosphate and carbonate compounds were employed in agriculture. A new group of insecticides, such as premethrin, cypermethrin, fenalerate, etc. which were effective at low doses came into being in the 1970s. The use of pesticides has helped in increasing agriculture production and also led to the development of resistance in pests, contamination of the environment and resurgence of many pests.

There are about 1000 agrochemicals in use in the world over. India accounts for about 3.7 per cent of the total world consumption. At present, our consumption is about 90,000 tonnes of plant protection chemicals. It comes to about 500 grams per ha compared to 10-12 kg/ha in Japan and 5 kg/ha in Europe. However, the use of pesticides in India is uneven like the fertilizers. While in cotton it is about 3 to 4 kg/ha, in pulses it comes to below 500 grams/ha.

Agricultural chemicals have become a major input in Indian agriculture with the increasing demand for food, feed and fibre. The pesticide consumption was about 2000 tonnes annually during the 1950s. India happens to be the second largest manufacturer of pesticides in Asia after Japan. It is also of interest to know that in spite of increased consumption of plant protection chemicals, the produce loss due to insects and pests increased by 5 times during the period from 1988 to 1995.

Increasing application of fertilizer also leads to increasing use of pesticides to control pests and diseases. The trend of increasing fertilizer use also compels the farmers to enhance the use of pesticides as well. the use of fertilizers in increasing amounts leads to growth of weeds and in the process of weedicide use many plants growing nearby also get killed, which reduces the biodiversity. Meanwhile, the weeds also develop resistance to herbicides and the quest to formulate even powerful herbicides begins.

Pesticide consumption in India from 1970-71 to 2001-02 is shown in Table.

Table. Pesticide consumption in India.

Sr. No.	Year	Consumption (.000 tonnes)
1	1970-71	24.32
2	1980-81	45.00
3	1990-91	75.00
4	1991-92	72.13
5	1992-93	70.79
6	1993-94	63.65
7	1994-95	61.36
8	1995-96	61.26
9	1196-97	61.26
10	1997-98	56.11
11	1998-99	52.44
12	1999-00	49.16
13	2000-01	46.20
14	2001-02	44.58

Consumption of pesticides increased from 24.32 thousand tonnes in 1970-71 to 75 thousand tonnes in 1990-91 and it slowed down during the subsequent period. Insects, pests and diseases like viral, bacterial and fungal affect the high yielding varieties of crops. Almost all pesticides are toxic in nature and pollute the environment leading to grave damage to ecology and human life itself. This indiscriminate use leaves toxic residues in foodgrans, fodder, vegetables, meat, milk, milk products, etc. besides in soil and water. High doses of pesticides severely affect the aquatic animals, fish and the wild life. Insects develop resistance to insecticides in crops like cotton and in turn force the farmers to the excessive use of them. Cases of pesticide poisoning and human and animal deaths are also reported. Pesticides irritate the skin and the respiratory system in the humans gets damaged. It was found that all water bodies like, rivers, canals, lakes, tanks and ponds and also the costal water were contaminated with high amounts of DDT, HCH and other organochlorine pesticides. River water is seen as more contaminated than other water sources.

Contamination of drinking water with DDT and HCH is reported from different states. Since the concentrations of contaminants are higher than MRL values fixed by the Environmental Protection Agency, the seriousness of the problem can be gauged.

Pesticides also contaminate animal feeds and fodder. Green fodder, paddy and wheat straw contain residues of DDT and HCH. Several studies have confirmed this trend. Milk and milk products are also affected by the pesticide use. Both bovine and human milk showed high levels of pesticide contamination. The sources of contamination of bovine milk are traced to the fodder and feed concentrates and in case of the human milk, the consumption of contaminated food by the lactating mothers is reported to be the reason.

Infant formula/baby milk powders also showed DDT and HCH contamination level ranging from 94 to 100 per cent. Butter and ghee, the other animal products revealed high contamination levels in many parts of the country. Cereals like wheat and rice were seen contaminated highly by pesticides like, DDT, HCH and malathion. The case of vegetable, vegetable oil, honey, fish etc. is also not different as they too have unacceptable high pesticide residue content levels.

The adverse effect of pesticide contamination on humans in India is understood from the study of dietary intake. Such studies, although a few in number, have confirmed high levels of pesticides contamination which come to more than 3 to 5 times than the agriculturally developed countries.

The daily intake of pesticide per individual is estimated to be about 0.51 milli-grams which is above the accepted level. The Indian Institute of Horticulture Research has reported contamination of 50 per cent of the fruits and vegetables sold in the Bangalore market with the residues of DDT and HCH. Use of herbicides over a period results in the shift of the weed flora. The weeds of minor importance, often, become major weeds. Repeated application of weedicides helps the development of resistance in weed at alarming proportions. The remedy recommended is rotation of herbicides or the use of other herbicides. Any way, the end result is contamination of ground water and soils inflicting damage on environment.

The number of herbicides registered in India comes to about 28 in 1997-98 which was only 10 before 10 years. This is often compared to about 300 herbicides available in the North America. There are only 10 herbicides manufactured in the country and the herbicides consumption was about 6000 Tonnes during 1994-95. It is reported that the level of herbicide use in rice, wheat, and tea in India is almost the same that of the world at large. Sugarcane, soyabean, groundnut, coffee, cotton, onion and potato are the other crops, which find widespread application of herbicides in India.

The contamination of water, air and soil with toxic synthetic fertilizers, pesticides and herbicides leads to increasing deaths of many creatures, and to human illness and mortality.

The end result is loss of biodiversity and natural harmony, increased expenditure to purify water, air, etc. The toxins in the food crops cannot be removed and the threat to human existence itself seems to be real.

The firms engaged in the manufacture and supply of agricultural inputs have a vested interest in keeping the input use increasing. Besides, they influence the government policy towards agriculture. '

Salinity and Water Logging

Water is one of the important inputs for the vigourous growth and high yields of crops. The modernisation of Indian agriculture has resulted in the increased use of irrigation water. The area under irrigation has grown substantially during the past three decades.

Table. Gross area under irrigation in India

Sr. No.	Year	Gross irrigated area (Million ha)
1	1970-71	38.18
2	1980-81	49.73
3	1990-91	62.47
4	1991-92	65.68
5	1992-93	66.76
6	1993-94	68.37
7	1994-95	70.65
8	1995-96	71.35
9	1996-97	73.25
10	1997-98	72.78

The gross irrigated area of 38.18 million ha in 1970-71 increased to 49.73 million ha in 1980-81 and the next decade ending 1990-91 saw this further rising substantially to about 62.47 million ha. It increased to 72.78 million ha during 1997-98. Heavy irrigation is necessary to get high production, as the new varieties cannot withstand water scarcity. This leads to salinity and water logging leaving the land uncultivable. Over exploitation of underground water is another effect. When water table falls, increasing energy will be required to lift water for irrigation.

Irrigation is necessary for the vigourous growth and high yields of crops in the modern method of cultivation. Many of the crops, particularly the rice and wheat high yielding varieties need more irrigation water than the traditional varieties. The area under irrigation in the country is only about 35 per cent and the remaining is still dependent on rains. So, there is a necessity to use irrigation water judiciously. Its excessive use results in severe ecological dangers like water logging of vast cultivated areas by seepages from canals. The loss of water through seepages and evaporation is estimated to be about 38 per cent. Flooding also results in run off and leaching losses of fertilizer nutrients, pesticides and soil particles. Excessive use of canal water makes the field vulnerable to soil erosion. The excessive irrigation in certain areas results in wastage as evident from the water logging of vast cultivated areas caused by the seepage from many virater sources.

Water logging is harmful to the soil. Seepage of canal water leads to salts present in the lowest layer of soil come up to the surface and the soil may turn alkaline or saline. Dams and multipurpose projects degrade the soil in the command area due to soil salinity and water logging. The chambel region in Rajasthan and Madhya Pradesh, the command area of the Bhakra Nangal, etc are the examples of water logging created by huge water irrigation projects. Crops irrigated by sewage water have adverse effects on the health of the human population consuming the produce. The workers work on these farms also face health hazards.

Depletion of Energy Resources

Chemical fertilizers, pesticides, herbicides, etc are manufactured using the non-renewable materials like the fossil fuels. The global demand for oil and natural gas is increasing and thus the price of the inputs to agriculture is bound to rise. India's petroleum resources, which presently meet only about 30-35 per cent of the consumption demand, are under pressure. Increasing demand for chemicals and energy in agriculture sector will have affects on our energy sources. The investments in agriculture have to be increased to meet the rising input costs and larger areas are brought under farming to earn profits. Large farms have to transport the produce to distant areas. Again, energy will be required for transportation, processing and packaging.

The rice-wheat cropping pattern and the cultivation of crops like sugarcane require high irrigation, which results in the depletion of water level. Singh and Singh found that the water level in the states of Punjab and Haryana had gone down by 0.3 to 1 m per annum during a period of 10 years due to the excessive use of water for paddy crop.

Input-Output Lmbalance

A crop, in its growth process, incorporates a part of the soil fertility into the parts of the plant. The roots remain in the soil. The leaves and stems are fed to the cattle/burnt as fuel/directly returned to the soil. The consumed part by cattle and human also go back to the soil. The practice of commercial farming leads to continuous export of the soil fertility to outside the farming areas as the organic matter leaves the locality. The soil nutrients in the form of farm produces continue to be exported. The import of chemical fertilizers cannot compensate the loss of soil nutrients through exports. The soil becomes powdery and gets eroded by wind or rain. If the harvests are exported from the country, the loss is higher.

The Earth can produce only a limited amount of biomass from a given area. If man tries to extracj: more, the system degenerates. If we realise that we live in a closed eco-system and act in tune, the resources can be used iri a sustainable way. The maintenance of a dynamic equilibrium balances the inputs and outputs in a region. If agriculture becomes unsustainable, no society can survive for a long time.

Expansion of Cultivated Area

Not only the intensive cultivation through the use of technological inputs, but also the extensive crop production through Increase in the area under cultivation has been an important aspect of modern agriculture seen in India. Increasingly areas under forests are brought under plough along with the marginal, sub-marginal and undulating land. The net sown area was 140 million ha in 1970-71 and stood at 142 million ha at the end of 1997-98.

Reduction in Genetic Diversity

The genetic base of crops is very important and a reduction of genetic diversity leads to the emergence of pests on a large scale. Farmers, in olden times, apart from' using the crop rotation methods to maintain the soil fertility also relied on the genetic means to increase crop production. Relying exclusively on nation's own reserves of fertility and immunology, the farming community by evolving trial and error methods discovered. hybrid varieties of crops by crossing the related strains. These crosses were from the same environment and no violence was used to separate them from nature by maintaining the ecological balance. The high yielding varieties of crops are the crosses from different environments and distantly related strains. The high yielding rice variety got by the crossing between the dwarf and non-dwarf varieties has major genetic weaknesses. The dwarf gene is susceptible to pest and viral attacks and the seed cannot manifest its potential without chemical fertilizer. Thus, an artificial environment has to be created for the growth of the crop.

Thus synthetic fertilizers supplant natural fertility, which results in larger population of pests. The new technology adopted then depends upon the replacement of the local/traditional varieties of seeds. But this results in the reduction of genetic diversity and increase in genetic erosion. These modern technologies are but the result of clever manipulations of nature's genes.

Low Productivity

The productivity of cereals, millets, oilseeds, pulses and plantation crops is very low in comparison with those in other countries in the world. This is in spite of our success in improving the quality of seeds and adoption of efficient technology. The impact of green revolution is showing signs of weakness and production appears to have decreased even after an increase in the inputs used.

The production of foodgrains in the country increased very substantially during 1960 to 1980 to reach 160 million Tonnes from 60 million tonnes. But the decade ending 1990 and 2000 did not witness such increases and the attainment of the targeted production of 240 million tonnes to meet the demand of the population by 2010 seems to be difficult. The reasons attributed to the low productivity are the drastic reduction in soil nutrients in the areas where fertilizer is used intensively in which the organic matter is not supplemented.

Benefits of Organic Farming

Organic agricultural practices are based on a maximum harmonious relationship with nature aiming at the non-destruction of the environment. The developed nations of the world are concerned about the spreading contamination of poisonous chemicals in food, feed, fodder and fibre. Naturally, organic farming system is looked upon as one of the means to remedy these maladies there. However, the major problem in India is the poor productivity of our soils because of the low level content of the organic matter.

The efficiency of the organic inputs in the promotion of productivity depends on the organic contents of the soil. There were many resemblances of organic farming principles in the traditional agriculture of India. But the former gives a more open and verifiable scientific foundation than the latter.

Healthy Foods

A study conducted in USA on the nutritional values of both organic and conventional foods found that consumption of the former is healthier. Apples, pears, potatoes, corn, wheat and baby foods were analysed to find out 'bad' elements such as aluminum, cadimum, lead and mercury and also 'good' elements like boron, calcium, iron, magnesium sellenium and zinc. The organic food, in general, had more than 20 per cent less of the bad elements and about 100 per cent more of the good elements.

Improvement in Soil Quality

Soil quality is the foundation on which organic farming is based. Efforts are directed to build and maintain the soil fertility through the farming practices. Multicropping, crop rotations, organic manures and pesticides, and minimum tillage are the methods employed for the purpose. Natural plant nutrients from green manures, farmyard manures, composts and plant residues build organic content in the soil.

Table. Organic carbon and available P contents

Year	Organic Available	Organic P (kg/ha)	Non-organic Available carbon(%)	Organic P (kg/ha) carbon(%)

June 1993	0.38	12.1	0.38	12.1
Feb 1994	0.40	12.6	0.36	12.0
Feb 1995	0.46	14.5	0.35	12.9
Feb 1996	0.52	15.0	0.38	12.0

It is reported that soil under organic farming conditions had lower bulk density, higher water holding capacity, higher microbial biomass carbon and nitrogen and higher soil respiration activities compared to the conventional farms. This indicates that sufficiently higher amounts of nutrients are made available to the crops due to enhanced microbial activity under organic farming.

Increased Crop Productivity and Income

Field trials of organic cotton at Nagpur revealed that during the conversion period, cotton yield was low compared to the conventional and integrated crop management. However, the yields of organic cotton started rising from third year. Cotton yields under organic, conventional and the mixed systems were 898, 623 and 710 kg/ha respectively at the end of the fourth year of the cultivation. The yield of soyabean under organic farming was also the highest compared to the other two systems.

The Central Institute for Cotton Research, Nagpur conducted a study of economics of cotton cultivation in Yavatmal district of Maharashta. The cost of cultivation of cotton was lower in the organic farming than in the modern system. The low costs were due to the non-use of fertilizers and chemical insecticides. As a result of the low yields during the conversion period, the net income from the organic farm was lesser than the conventional farm. But the yield under organic method increased progressively equalling it to that of the conventional system by the sixth year.

The input costs were low under organic farming and with a 20 per cent of premium prices of output, the net income increased progressively from fourth year under organic fanning. The appreciation of net income from organic cotton cultivation by the sixth year was 80 per cent over the conventional crop.

Results reported from 1050 field demonstration cum trials under the National Project on Development and Use of Biofertilizers in different parts of the country show* an increase of 4 per cent in yield in plantation crops, 7 per cent in fruit crops, 9 per cent in wheat and sugarcane, 10 per cent in millet and vegetable, 11 per cent in fibre, condiments and spice crops, 14 per cent in oilseeds and flowers and 15 per cent in tobacco.

A study of 100 farmers in Himachal Pradesh during a period of 3 years found that the total cost of production of maize and wheat was lower under organic farming and the net income was 2 to 3 times higher. Both productivity and premium prices contributed to the increased profitability. Another study of 100 farmers of organic and conventional methods in five districts of

Karnataka indicated that the cost of organic farming was lower by 80 per cent than that of the conventional one.

Table. The Cost Benefit Ratios Described for Various Crops

Crops	Organic	Inorganic
Groundnut	1:1.26	1:1.31
Jowar	1:1.36	1:1.28
Cotton	1:1.34	1:1.24
Coconut	1:1.70	1:1.31
Banana	1:3.66	1:2.82

Low Incidence of Pests

The study of the effectiveness of organic cotton cultivation on pests at the farm of Central Institute for Cotton Research, Nagpur revealed that the mean monthly counts of eggs, larva and adults of American BoUworm were far lesser under organic farming than under the conventional method. Bio-control methods like the neem based pesticides to Ti-ichoderma are available in the country. Indigenous technological products such as Panchagavya which was experimented at the University of Agricultural Sciences, Bangalore found to control effectively wilt disease in tomato.

Employment Opportunities

Organic farming requires more labour input than the conventional farming system. Thus, India which has a very large amount of labour unemployment and under employment will find organic farming an attraction. Moreover, the problem of periodical unemployment will also get mitigated because of the diversification of the crops with their different planting and harvesting schedules resulting in the requirement of a relatively high labour input.

Indirect Benefits

Several indirect benefits from organic farming are available to both the farmers and consumers. While the consumers get healthy foods with better palatability and taste and nutritive values, the farmers are indirectly benefited from healthy soils and farm production environment. Eco-tourism is increasingly becoming popular and organic farms have turned into such favourite spots in countries like Italy. Protection of the ecosystem, flora, fauna and increased biodiversity and the resulting benefits to all human and living things are great advantages of organic farming which are yet to be properly accounted for.

METHODS OF ORGANIC FARMING

As you read and learn more about organic farming, you may encounter some differences in approach, style or what might be called philosophy.

While there have likely always been some farmers who have grown crops without fertilizers or pesticides, these are several examples of people who proposed or popularised a particular method. Please do not let this presentation format of separating them out give the impression that one or the other method must be selected. The goal is just to help give some history, background, and names to help you learn more.

BIODYNAMIC

Perhaps the oldest (this century) recognised method of organic gardening or farming is referred to as the Biodynamic Method. Rudolph Steiner developed the method in Germany during the 1920's and "his basic theory was that nature is a mystical, spiritual thing, and to garden it well, one must treat it as a whole entity and system, both physical and spiritual". Farming practices are related to lunar and planetary phases. Plant and mineral based preparations are often sprayed on crop plants to provide necessary nutrients.

THE RODALES

J.I. Rodale searched for alternative agriculture methods in the early 1940's. He was not from a farming background but wanted to grow more of his own food.

The recommendations of the day to treat with chemicals that he saw as potential poisons did not make sense to him. He learned about the composting methods of Sir Albert Howard in India (*An Agricultural Testament*), the work of Lady Eve Balfour (*The Living Soil*), and F. King (*Farmers of Forty Centuries*) who studied growing systems in Asia. Rodale developed an emphasis on human health and how organic gardening through reduced exposure to pesticides and production of nutrient rich vegetables was related to human health. A major emphasis was placed on the production and use of compost to enrich the soil. The work was continued by his son, Robert Rodale. The family developed the Rodale Institute, Organic Farming and Gardening Magazine, and Rodale Press.

BIOINTENSIVE

The Biointensive Method is mostly outlined in the book *How to Grow More Vegetables Than You Ever Thought Possible on Less Land Than You Can Imagine* by John Jeavons. High organic matter, growing organic matter specifically for soil building, deep cultivation or double digging, and extremely high yields per unit area of production are some key principles. As reported in the book mentioned above, the techniques were originally outlined by Alan Chadwick, an English horticulturist who combined French Intensive gardening methods such as double digging with Biodynamic methods. Extremely high yields have been achieved on small plots of land around the world.

NATURAL FARMING

The Natural Farming method of gardening or farming is based on building soil with regular surface applications of organic matter and little or no cultivation. Organic matter is allowed to slowly decompose on the surface as opposed to composting in piles. Weeds are controlled by applications of more mulch or organic matter. This method was popularised by Masanobu Fukuoka in Japan.

NO WORK METHOD

This method is very similar to the Natural Farming method. Ruth Stout published a book in 1971 called The Ruth Stout No-Work Garden Book" (Rodale Press, Emmaus, Penna, 1971). A recent book titled Gardening Without Work : For the Aging, the Busy and the Indolent, 1998 is likely a reprinting of the information. The emphasis in on lots of organic matter for mulch to control weeds and provide nutrients. Worms and other organisms are allowed to do the work of getting nutrients to the roots.

PERMACULTURE

A broader, more holistic method know as Permaculture has been popularised by Bill Mollison and others. It appears to have originated in Australia or New Zealand. Permaculture goes even farther to evaluate and take into account the entire landscape and physical features of a growing area.

Plant and animal raising are considered together to make the best use of the land. Sustainability of practices and design and care for the earth are key components of Permaculture. In some ways it appears to be a process of observing nature and methods of indigenous people and perhaps making them formal or complicated or academic.

VEGANICS

Some people have chosen to use the term *veganics* as opposed to *organics*. The main difference is basing nutrient management on only plant derived nutrients and organic matter as opposed to making use of manures and other animal products. This is not a widely used term but is does point out that there are different degrees of organic farming/gardening and what people consider important. Regardless of what you call it, the ideas have common threads. Organic gardening is gardening in a way that considers the whole picture of the environment and how we fit into the existing cycles. Gardening with nature is a common sense way of putting it. There are many other methods that could be considered, including the French Intensive gardening methods and the old and widely used method of China and Southeast Asia based on intensive cultivation of diversified plots. We could also consider the work of Wes Jackson and other at The Land Institute in Kansas that are working on methods to use more perennial plants for staples such as wheat, rice and corn. Hopefully there are enough choices and ideas here that you won't get the impression that there is only one way or one recipe to follow. Just farm with nature.

VARIETIES FOR ORGANIC FARMING

Widespread use of improved rice varieties since the 1960s has reduced food prices, for the poor and prevented millions of cases of childhood malnutrition. Without the development of the high yielding varieties, prices for developing country consumers would likely be as much as 40 per cent higher than they are today. The new varieties have also reduced costly food imports by almost 8 per cent and have eliminated the need to convert millions of hectares of forestland to agricultural uses as would have otherwise been required had yields remained at 1960 levels. The availability of high yielding varieties has prompted many of the world's poorest countries to invest in plant breeding programmes and produce varieties suited to local environments and markets. The new plant types have also prompted massive government investments in agricultural infrastructure such as irrigation and fertilizer delivery systems. In spite of the advantages of high yielding varieties, it should not be forgotten that the adoption of high yielding has led to the substitution of a large quantity of species for only a few and uniform varieties from a genetic point of view, which has caused a significant reduction in the genetic inheritance of cultivated species. Many agricultural species, varieties and breeds which have played an important role in the human diet and traditional cultures have practically disappeared over the last century.

In the last decade, the adoption of organic agriculture has indirectly established a rescue process of species, varieties and breeds threatened by under-use or extinction. Stronger collaboration has been evident among movements aiming to defend biodiversity and the organic agriculture movement. This is especially the case now that there is interest in traditional, speciality and organic products. For the rescue of varieties threatened by extinction, the development of a market is fundamental and it is here that organic agriculture plays an important role as the price premium gives an additional value to the product. The restoration and enhancement of under-utilized species and varieties has been motivated by a food demand concerned with health and culinary traditions. Organic agriculture has allowed the maintenance and improvement of species and varieties that otherwise would suffer strong genetic erosion or extinction.

ORGANIC AGRICULTURE

Organic systems avoid the use of synthetic fertilizers, pesticides, and growth regulators. Instead they rely on crop rotations, crop residues, animal manures, legumes, green manures, off-farm wastes, mechanical cultivation, mineral-bearing rocks, and biological pest control to maintain soil health, supply plant nutrients, and minimize insects, weeds, and other pests.

SELECTION OF RICE VARIETIES FOR ORGANIC FARMING

In general, rice planting dates, seeding rates, preferred varieties, and harvesting methods vary among regions, but they are largely the same for conventional and organic systems While choosing varieties for organic farming, the special considerations relevant to *organic* rice production has to be duly recognized. Weed control, management of soil fertility and minimizing pests and diseases are the principal challenges associated with organic rice production and according to the needs and requirements in each, varieties have to be chosen.

WEED CONTROL

Weeds are a problem in both wetland rice as well as up lands, but it is more hazardous under up land conditions. Direct seeding of rice is receiving much attention in recent times since it reducesproduction costs. As cultural practices for rice shift from transplanted to direct seeded rice, weed problems will increase because rice and weeds can emerge together.

Many weeds of irrigated ecosystems like Barnyard grass can become more important/harmful in direct seeded rice since they are adapted for better growth under dry than wet conditions. While choosing varieties for upland rice, care should be taken to select varieties with initial vigour and increased growth rate so that they can compete with weeds and come up well. Short duration varieties are preferred for direct seeding in uplands, since they have higher growth rate compared to medium or long duration varieties.

Girija *et al.* while evaluating upland varieties for their weed competitiveness at Regional Agricultural Research Station, Pattambi, noticed distinct difference in the weed suppression ability of the cultivars tested.

Based on this, they classified the rice varieties into three groups viz.

1. Varieties, which can completely smother the weeds,
2. Varieties that show moderate tolerance and
3. Varieties that are smothered by competing weeds.

Improved traditional varieties like PTB 28, PTB 29 and PTB 30 and land races like Karanellu belong to the first group. The improved upland varieties could grow fast and produce large no. of tillers, there by preventing the growth of weeds in between. Land races like Karanellu had long and droopy leaves, which prevented light interception by weeds there by reducing their growth. Research is underway to transfer the weed competitiveness of these varieties to high yielding varieties.

In transplanted rice in lowlands, crop rotations, land leveling, seedbed preparation, water management, and rotary hoeing are the primary weed-control practices followed to take care of the weeds. Here also, varieties having natural weed fighting capacity could be chosen to reduce the hazards from high weed growth. Use of allelopathic rice varieties having activity against a broad spectrum of target weeds could be a desirable choice. Cultivars that show allelopathic potential against important rice weeds have been identified in many countries *viz*., United States, Japan, Egypt and the Philippines. In the United States, evaluated more than 10000 accessions of rice for allelopathic effects against aquatic weeds like duck salad and red stem and could identify around 190 promising accessions. This included rice varieties from many countries including India. Allelopathic cultivars that strongly inhibit root elongation of barnyard grass but weakly affect the shoot have been identified by Olofsdotter. In Egypt, identified Indian rice varieties that expressed allelopathic effects on Echinocloa cruss-galli and Cyperus difformis. L. at 3-4-leaf stage by inhibiting root development and emergence of the first or second leaf of both weeds. Kim and Shin, 1998 identified promising rice germplasm for allelopathic activity in Korea and new cultivars are being generated using them.

SOIL FERTILITY

Maintaining soil fertility in organic cropping typically involves some combination of crop rotation with deep-rooted legume crops or green manure/ cover crops, and applying rock minerals, animal manures, composts, and other approved organic amendments. Green manures have proven their positive influence in enhancing rice yields. Leguminous green-manure crops can supply 30 to 50 per cent of the nitrogen needs of high yielding rice varieties. The availability of green-manure nitrogen depends on the quantity, quality, and type of green-manure crop; the time and method of application; soil fertility; and cropping method.

It has been observed that green manuring with Sunnhemp or Cowpea grown during summer months and incorporated in soil before transplanting rice could save about 60 kg N/ha for rice crop. Recent studies on economizing chemical fertilizers through organic manures in selected cropping systems under All India Co-ordinated Agronomic Research Project have shown that 25-50% N requirement of rice during Kharif season could be met through organic sources without any adverse effect on rice productivity. Yield response of rice varieties to organics indicated varietal differences to organic farming. Some varieties responded more to organics than to chemical fertilizers. Similarly, certain rice varieties responded to chemical fertilizers than to organics.

Emission of trace gases especially Methane from flooded rice fields has become a serious concern world over. Methane emission from lowland rice has been identified as resulting from organic matter fermentation and factors like incorporation of crop residues like rice straw can enhance the process. Varieties with low methane emission potential could reduce 'hazards resulting from high methane emission and could be more rewarding under organic farming.

INSECTS AND DISEASES

Rice, both rain fed and irrigated is infested by a variety of pests and diseases. Intensive cropping of modern varieties under high inputs provided a favourable environment for many pests as a limited number of modern varieties began to be widely cultivated, the populations and economic importance of their associated pests increased and new pathogen strains and insect biotypes developed. The major insect pe.sts *viz.*, Plant hoppers gall midge, stem borer, leaf rollers, rice bug, along with minor pests like case worm, thrips, ear cutting caterpillar, whorl maggot etc., cause serious damage to the rice crop in the tropics. Similarly, fungal diseases (leaf blast, sheath blight, brown spot, sheath rot, bacterial diseases (bacterial blight) and viral diseases (Tungro Virus, Grassy Stunt Virus) etc., can also devastate the crop. Excessive nitrogen levels are rarely a problem in organic production, since use of synthetic fertilizers is avoided, but timely control measures are a must to keep the damage under economic threshold levels. Timely planting, variety selection, and cultural practices to suppress weeds and encourage dense stands of rice will help control most of the biotic stresses.

Use of resistant varieties can prove to be consistently and significantly more productive under high pest or disease pressure. Varieties with moderate resistance are a better choice since they allow pest populations to be maintained at levels that do not result in significant damage. Varieties with strong resistance reduce the populations of natural enemies that feed on the pest and could ultimately lead to resurgence of the pest. List of high yielding varieties combining resistance to various biotic stresses released by the Central Seed Subcommittee on Crop Standards, Notification and Release of Varieties. The different states have also released a number of varieties resistant to various biotic stresses.

Selecting varieties possessing multiple resistances could make the programme more effective. Again, the economic loss associated with each stress should be assessed before selecting a variety and high priority should be assigned to varieties resistant to a particular stress for which biological control measures are not available, and low priority could be given for varieties resistant to such stresses where other low cost control methods are available. For eg. Use of varieties resistant to Bph and gall midge and adopting biological control measures for stem borer and leaf roller could take care of all the above pests. Similarly, selecting a blast resistant variety along with biological control measure for sheath blight could be a more effective method for controlling the two diseases rather than going for varieties resistant to both.

NUTRIENT MANAGEMENT IN ORGANIC FARMING

Organic farming is often understood as a form of agriculture with use of only organic inputs for the supply of nutrients and management of pests and diseases. In fact, it is a specialized form of diversified agriculture, wherein problems of farming are managed using local resources alone. The term organic does not explicitly mean the type of inputs used; rather it refers to the concept of farm as an organism. Often, organic agriculture has been criticized on the grounds that with organic inputs alone, farm productivity and profitability might not be improved because the availability of organic sources is highly restricted7. True, organic resources availability is limited; but under conditions of soil constraints and climate beggaries, organic inputs use has proved more profitable compared to agrochemicals.

Organic farming systems rely on the management of soil organic matter to enhance the chemical, biological and physical properties of the soil. One of the basic principles of soil fertility management in organic systems is that plant nutrition depends on 'biologically-derived nutrients' instead of using readily soluble forms of nutrients; less available forms of nutrients such as those in bulky organic materials are used. This requires release of nutrients to the plant via the activity of soil microbes and soil animals. Improved soil biological activity is also known to play a key role in suppressing weeds, pests and diseases.

Animal dung, crop residues, green manure, biofertilizers and bio-solids from agro-industries and food processing wastes are some of the potential sources of nutrients of organic farming. While animal dung has competitive uses as fuel, it is extensively used in the form of farmyard manure. Development of several compost production technologies like vermicomposting, phosphocomposting, N-enriched phosphocomposting, etc. improves the quality of composts through enrichment with nutrient-bearing minerals and other additives. These manures have the capacity to fulfill nutrient demand of crops adequately and promote the activity of beneficial macro- and micro-flora in the soil.

There are several doubts in the minds of not only farmers, but also scientists about whether it is possible to supply the minimum required nutrients to crops through organic sources alone. Even if it is possible, how are we going to mobilize the organic matter? At this juncture, it is neither advisable nor feasible to recommend the switchover from fertilizer use to organic manure under all agro-ecosystems. Presently, only 30% of our total cultivable areas has irrigation facilities where agrochemicals use is higher compared to rain-fed zones. It is here that ingenuity and efforts are required to increase crop productivity and farm production despite recurrence of environmental constraints of drought and water scarcity.

The basic requirement in organic farming is to increase input use efficiency at each step of the farm operations. This is achieved partly through reducing losses and adoption of new technologies for enrichment of nutrient content in manure. Technologies to enrich the nutrient supply potential from manure, including farmyard manure three to four times are being widely used in organic farms. Around 600 to 700 million tonnes (mt) of agricultural waste is available in the country every year, but most of it is not used properly. We must convert our filth into wealth by mobilizing all the biomass in the rural and urban areas into bioenergy to supply required nutrients to our starved soil and fuel to farmers. India produces about 1800 mt of animal dung per annum. Even if two-thirds of the dung is used for biogas generation, it is expected to yield biogas not less than 120 m m3 per day. In addition, the manure produced would be about 440 mt per year, which is equivalent to 2.90 mt N, 2.75 mt P2O5 and 1.89 mt K2O.

Organic farms and food production systems are quite distinct from conventional farms in terms of nutrient management strategies. Organic systems adopt management options with the primary aim to develop whole farms, like a living organism with balanced growth, in both crops and livestock holding. Thus nutrient cycle is closed as far as possible. Only nutrients in the form of food are exported out of the farm. Crop residues burning is prohibited; so also the unscientific storage of animal wastes and its application in the fields. It is, therefore, considered more environment friendly and sustainable than the conventional system. Farm conversion from high-input, chemical-based system to organic system is designed after undertaking a constraint analysis for the farm with the primary aim to take advantage of local conditions and their interactions with farm activities, climate, soil and environment, so as to achieve closed nutrient cycles with less dependence on off-farm inputs. This implies that the only nutrients leaving the farm unit are those for human consumption.

Crop rotations and varieties are selected to suit local conditions having the potential to sufficiently balance the nitrogen demand of crops. Requirements for phosphorus, sulphur and micronutrients are met with local, preferably renewable resources. Organic agriculture is, therefore, often termed as knowledge-based rather than input-based agriculture. Furthermore, organic farms aim to optimize the crop productivity under a given set of farm conditions. This is in contrast to concept of yield maximization through the intensive use of agrochemicals, irrigation water and other off-farm inputs. There are ample evidences to show that agrochemical-based, high-input agriculture is not sustainable for long periods due to gradual decline in factor productivity, with adverse impact on soil health and quality.

ORGANIC FARMING IS BETTER FOR THE ENVIRONMENT

As an ecologist by training, this myth bothers me the most of all three. People seem to believe they're doing the world a favor by eating organic. The simple fact is that they're not – at least the issue is not that cut and dry.

Yes, organic farming practices use less synthetic pesticides which have been found to be ecologically damaging. But factory organic farms use their own barrage of chemicals that are still ecologically damaging, and refuse to endorse technologies that might reduce or eliminate the use of these all together. Take, for example, organic farming's adamant stance against genetically modified organisms (GMOs).

GMOs have the potential to up crop yields, increase nutritious value, and generally improve farming practices while reducing synthetic chemical use – which is exactly what organic farming seeks to do. As we speak, there are sweet potatoes are being engineered to be resistant to a virus that currently decimates the African harvest every year, which could feed millions in some of the poorest nations in the world15. Scientists have created carrots high in calcium to fight osteoperosis, and tomatoes high in antioxidants. Almost as important as what we can put into a plant is what we can take out; potatoes are being modified so that they do not produce high concentrations of toxic glycoalkaloids, and nuts are being engineered to lack the proteins which cause allergic reactions in most people. Perhaps even more amazingly, bananas are being engineered to produce vaccines against hepatitis B, allowing vaccination to occur where its otherwise too expensive or difficult to be administered. The benefits these plants could provide to human beings all over the planet are astronomical.

Yet organic proponents refuse to even give GMOs a chance, even to the point of hypocrisy. For example, organic farmers apply *Bacillus thuringiensis* (Bt) toxin (a small insecticidal protein from soil bacteria) unabashedly across their crops every year, as they have for decades. It's one of the most widely used organic pesticides by organic farmers. Yet when genetic engineering is used to place the gene encoding the Bt toxin into a plant's genome, the resulting GM plants are vilified by the very people willing to liberally spray the exact same toxin that the gene encodes for over the exact same species of plant. Ecologically, the GMO is a far better solution, as it reduces the amount of toxin being used and thus leeching into the surrounding landscape and waterways. Other GMOs have similar goals, like making food plants flood-tolerant so occasional flooding can replace herbicide use as a means of killing weeds. If the goal is protect the environment, why not incorporate the newest technologies which help us do so?

But the real reason organic farming isn't more green than conventional is that while it might be better for local environments on the small scale, organic farms produce far less food per unit land than conventional ones. Organic farms produce around 80% that what the same size conventional farm produces16 (some studies place organic yields below 50% those of conventional farms!).

Right now, roughly 800 million people suffer from hunger and malnutrition, and about 16 million of those will die from it17. If we were to switch to entirely organic farming, the number of people suffering would jump by 1.3 billion, assuming we use the same amount of land that we're using now. Unfortunately, what's far more likely is that switches to organic farming will result in the creation of new farms via the destruction of currently untouched habitats, thus plowing over the little wild habitat left for many threatened and endangered species.

Already, we have cleared more than 35% of the Earth's ice-free land surface for agriculture, an area 60 times larger than the combined area of all the world's cities and suburbs. Since the last ice age, nothing has been more disruptive to the planet's ecosystem and its inhabitants than agriculture. What will happen to what's left of our planet's wildlife habitats if we need to mow down another 20% or more of the world's ice-free land to accommodate for organic methods?

The unfortunate truth is that until organic farming can rival the production output of conventional farming, its ecological cost due to the need for space is devastating. As bad as any of the pesticides and fertilizers polluting the world's waterways from conventional agriculture are, it's a far better ecological situation than destroying those key habitats altogether. That's not to say that there's no hope for organic farming; better technology could overcome the production gap, allowing organic methods to produce on par with conventional agriculture. If that does occur, then organic agriculture becomes a lot more ecologically sustainable. On the small scale, particularly in areas where food surpluses already occur, organic farming could be beneficial, but presuming it's the end all be all of sustainable agriculture is a mistake

THE IMPACT OF ORGANIC AGRICULTURE ON FOOD QUALITY

During last decades the consumer trust in food quality has drastically decreased, mainly because of the growing ecological awareness and several food scandals like Bovine Spongiform Encephalopathy (BSE), dioxins and bacterial contamination. It has been found that intensive conventional agriculture could introduce contaminants into food chain, first of all nitrates and nitrosamines, residues of pesticides, antibiotics and growth hormones. Consumers started to look for safer and better controlled foods, produced in more environmentally friendly, authentic and local system. Organically produced foods are widely believed to satisfy the above demands, providing better environment and higher nutritive values. Several research studies conducted in many European countries have partly confirmed this opinion.

Organic crops contain fewer nitrates and nitrites and fewer residues of pesticides than conventional ones. They contain as a rule more dry matter, more vitamin C and B-group vitamins, more phenolic compounds, more exogenous indispensable amino acids and more total sugars; however the level of â carotene is often higher in conventional plant products. Organic crops contain statistically more iron, magnesium and phosphorus, and they have usually better sensory quality. Vegetables, potatoes and fruits from organic production show better storage quality during winter keeping. Farm animals from organic herds show less metabolic diseases like ketosis, lipidosis, arthritis, mastitis and milk fever. Small experimental mammals (rats, rabbits) fed organically grown feed show better health and fertility parameters.

However, there are also some negatives: plants cultivated in organic system have as a rule 20 % lower yield than conventionally produced crops. Milk and meat yield is also lower in organic animal production, partly because parasitic afflictions are more frequent. Several important problems need to be investigated and settled in coming years: environmental contamination of the organic crops (heavy metals, polychlorinated biphenyls (PCBs), dioxins, and aromatic hydrocarbons), bacterial and fungi contamination (Salmonella, Campylobacter, mycotoxins). Last but not least, the impact of the organic food consumption on human health and well being still remains unknown and needs explanation.

9

Mineralization of Organic Matter

MINERALIZATION OF THE ORGANIC MATTER (OM)

A fundamental hypothesis of diagenetic models such as those used by Berner (1977) or Rabouille and Gaillard (1990) is the existence of a steady state within the pore water composition. A very active bioturbation is not compatible with this hypothesis, but, the main geochemical processes being now identified, it is possible to quantify the amount of OM mineralized at station 8 through the observed concentration profiles and a simple stoichiometric model.

THE OXIDATION RÉACTIONS

At the very top of the sedimentary column (a few millimeters below the interface) dissolved oxygen may diffuse from the water column and part of the Organic Matter (OM) was oxidized according to the following reaction:

$$``CH_2O''^{*} + O_2 \rightarrow CO_2 + H_2O$$

The OM can be oxidized by sulfate, in a first approximation,according to the following reaction:

$$``CH_2O'' + \tfrac{1}{2}\, SO_4^{=} + H^{+} \rightarrow \tfrac{1}{2}\, H_2S + CO_2 + H_2O$$

* "CH_2O" is OM

Rewritten more accurately by taking into account the contributions of nitrogen and phosphorus :

- *N and P*
- In the oxygenated 1 cm top layer :

$$(CH_2O)_x\,(NH_3)_y\,H_3PO_4 + xO_2 \rightarrow xCO_2 + yNH_3 + H_3PO_4 + xH_2O$$

- And below :

$$(CH_2O)_x\,(NH_3)_y\,H_3PO_4 + x/2\ SO_4^{=} + xH^{+} \rightarrow x/2\ H_2S + xCO_2 + yNH_3 + H_3PO_4 + xH_2O$$

- For the whole profile :

$$CaCO_3 + CO_2 + H_2O \leftrightarrow Ca^{++} + 2HCO_3^{-}$$

- *For the total dissolved CO_2*
- Within the oxygenated layer we can write :

$$\Delta\Sigma CO_2 = \Delta O_2 + \Delta Ca$$

- And below :

$$\Delta\Sigma CO_2 = \Delta H_2S + \Delta Ca$$

- *For the alcalinity :*
- Within the oxygenated layer we can write :

$$\Delta\Delta lk = (1\text{-}y)/x\ \Delta O_2 + 2\Delta Ca$$

The borate contribution is supposed to be constant and the contribution of $H_3SiO_4^-$ is neglected.

- And below :

$$\Delta Alk = 2(1\text{-}y\text{-}1)/x\ \Delta H_2S + 2\Delta Ca$$

In order to improve the calculation of Alk, the profile of dissolved sulfide (H2S) has been fitted with a polynomial function. Therefore, it is possible to get a good estimation of H_2S for each depth incrementz.

THEORICAL PROFILE OF ALKALINITY

The stoichiometric modeling will be applied only to generate a theoretical alkalinity profile.

The experimental alkalinity profile is presented on the fig. for station 8 together with the calculated one. The agreement between observed and calculated values is fairly good when the C:N ratio is set to 9.6. This high ratio compared to the Redfield et al. (1963) ratio (6.6), reflects that The OM oxidized within the sediment is a mixture containing phytoplankton and an other type ofOM with a lower C:N ratio which could be a contribution of dead benthic material.

DIAGENIC REACTIONS AND CHEMICAL EQUILIBRIUM

Quantitative estimation of the diagenetic reactions

We can estimate the amount of OM mineralized by the two main oxidation reactions.

- *In the upper centimeter:*

$\Delta OM/\Delta O_2 = 1/x$. If $x = 106$ and $O_2 = 187$ µM, then OM = 1.8 µM. In the top centimeter of the sediment the porosity= 0.62. Assuming a density of 2.5 g/cm^3 for coral sand and 1.025 g/cm^3 for sea water, then we find that 2.5 mg of OM is oxidized per kg of total sediment (solid + pore water).

Below and down to 34 cm below the interface:

$\Delta OM/\Delta O_2 = 2/x$. Si $x = 106$ et $H_2S = 116$ µM, then OM = 2.2 µM. With an average porosity value = 0.57 (between 1 and 34 cm below the interface) the same calculation gives 2.7 mg of OM per kg of total sediment.

For the whole sediment

The extent of the mineralization processes is restricted to 5.2 mg of organic material per kg of sediment (5.2 ppm). This value seems very low compared to the results obtained by the litterature. Our result emphasizes the oligotrophic status of the global lagoonal ecosystem.

Chemical equilibrium between the aragonitic coral sand and the pore water

To determine if the coupled reaction of dissolution-precipitation of the calcium carbonate occurs within equilibrium conditions, a saturation index (SI) have been calculated with respect to the most probable carbonate phase, the aragonite and alternatively for calcite.

$$SI = [Ca]\ [CO_3]/\ Ks$$

[Ca] and $[CO_3]$ represent the total dissolved calcium and carbonate respectively,

Ks = stoichiometric constant of solubility

- The values of Ks are 6.34 10-7 and 4.23 10-7 for aragonite and calcite respectively, at T = 301 K and S = 35 psu.
- [Ca] is directly given by analysis.
- $[CO_3]$ must be calculated from the values of the carbonate alkalinity and the pH:

$$[CO_3] = Alk_c/\ (\ 2 + 10^{\ (pK2 - pH)}\ )$$

pK_2 is the second dissociation constant of H_2CO_3.

With a value of SI very close to 1, an equilibrium between the pore water and the sediment can be inferred. With SI much greater than 1, the solution is supersaturated with respect to a solid phase which can precipitate. The opposite situation, when SI is smaller than 1, reflects an under-saturation where the solid phase considered can dissolve.

The figure shows that the whole pore water profile is supersaturated with respect to both carbonates. The highest supersaturation occurs within the first top centimeter (SI = 5.5 and 8.2 for aragonite and calcite respectively at the SWI). This confirms the hypothesis that the oxidation of organic matter by dissolved oxygen leads to the dissolution of coral sand, while this dissolution does not appear on the calcium profile. Downward, supersaturation decreases but remains close to SI = 2 for aragonite. Then, when oxidation of the OM is controlled by sulfate, the carbonate phase precipitates and refrains any pH increase.

MINERALIZATION (SOIL SCIENCE)

Mineralization in soil science is decomposition or oxidation of the chemical compounds in organic matter into plant-accessible forms. Mineralization is the opposite ofimmobilization.

SOIL ORGANIC MATTER

Soil organic matter makes up only a few percent of most soils, but it has a great deal of influence on soil properties, and in turn, agricultural productivity.

What does it do? The list of soil properties affected by soil organic matter is long. It includes: aggregate stability (how well tiny clumps of soil hold together, which affects soil structure); cation exchange capacity (the ability of soil to hold onto positively charged nutrients for plant growth), nutrient release rate by mineralization (how much nitrogen, phosphorus and other elements are given off by microbial activity), and water-holding capacity.

How much is there? The amount of soil organic matter in a particular location is primarily due to natural factors like temperature (cool locations accumulate more organic matter), soil texture (clayey and silty soils tend to have more organic matter than sandy soils), and the drainage (poor drainage promotes soil organic matter build up). However, management also affects soil organic matter level, and it is not unusual for that level to decline over time in cultivated fields. Frequent tillage, periods of bare ground, and removal of crop residues all contribute to reductions in soil organic matter.

What is it? Soil organic matter is made up of plant and animal residues in different stages of decomposition, cells of soil microorganisms, and substances that are so well-decomposed it's impossible to tell what they were to begin with.

Living organisms are also considered to be part of soil organic matter, and they play a big role in contributing organic residues to the soil and in formation of more stable types of organic matter. Plant roots and various soil animals (rodents, earthworms, mites, etc.) all provide organic materials to the soil that eventually become part of the soil organic matter cycle. There are four main processes in that cycle, and all of them rely on soil microbes: decomposition of organic residues, release of nutrients (mineralization), release of carbon dioxide (respiration), and transfer of carbon from one soil organic matter 'pool' to another.

Not just one kind of organic matter. In addition to organic matter that is alive, there are three types, or pools, of "dead" soil organic matter: active, slow, and passive. These are determined by the time it takes for them to completely decompose.

Active soil organic matter is primarily made up of fresh plant and animal residues that break down in a very short time, from a few weeks to a few years. This kind of organic matter is associated with a lot of biological activity. Passive soil organic matter, also known as humus, is not biologically active, meaning it provides very little food for soil organisms. It may take hundreds or even thousands of years to fully decompose! Slow soil organic matter is somewhere in between active and passive soil organic matter. It consists primarily of detritus, partially broken down cells and tissues that are only gradually decomposing. Slow soil organic matter is somewhat resistant to decay and may take a few years to a few decades to completely break down.

What is Humus? Hummus (with two m's) is a middle-eastern food made from chick peas. Humus (one m) is the passive fraction of soil organic matter. It is a dark, complex mixture of organic substances that have been significantly modified from their original form over time, and it also contains other substances that have been synthesized by soil organisms. Usually, humus represents the majority of total soil organic matter, and it is relatively stable over time.

Humus has a lot to do with the ability of a soil to retain nutrients and water. Humus also supplies organic chemicals to the soil solution that can serve as chelates, which can hang onto trace elements and increase their availability to plants.

What is active soil organic matter? Active soil organic matter is 'fresh meat' to microbes. It is the readily digestible and easily decomposed portion of fresh organic (meaning carbon-containing) residues. Active soil organic matter plays a very different role than passive organic matter does. As it is decomposed by soil organisms it helps stabilize soil aggregates, it releases nutrients by mineralization, and it provides food for microbial activity, which can lead to suppression of plant diseases and enhanced plant growth.

The amount of active organic matter in the soil can change quickly, in just a year or two, and it's highly influenced by soil management practices. To maintain the same level of active soil organic matter requires a constant supply of fresh organic materials, usually from growing plants. Crop roots, crop residues and cover crops all contribute to active organic matter. Soil must also be managed to minimize the loss of organic matter through oxidation (from aggressive tillage) and erosion (from ground left bare).

The amount of active soil organic matter and the proportion it makes up of total soil organic matter are good indicators of soil health. Unfortunately, most soil tests don't differentiate between the different forms of soil organic matter; they just report the percent of soil that is total organic matter. As a result, farmers aren't able to determine what is going on with their soil organic matter pools, and they can't monitor how their management practices are affecting active soil organic matter levels.

The good news is that soil scientists are working to develop affordable tests active soil organic matter, and these should be commercially available in the near future.

SOIL ORGANIC MATTER NUTRIENT MINERALISATION

US data is demonstrating that soil organic matter mineralisation plus biomass respiration can contribute significant amounts of plant available nutrients for crops and pastures. This review by USDA soil scientist Mike

Kucera, suggests high microbial biomass soils (particularly with high active fraction or labile organic matter) may not need any fertiliser nitrogen to achieve optimum crop yields. Requirements for fertiliser phosphorus are also reduced. Patrick Francis

Soil organic matter (SOM) is the organic component of soil, consisting of three primary parts including small (fresh) plant residues and small living soil organisms, decomposing (active) organic matter, and stable organic matter (humus).

Soil organic matter serves as a reservoir of nutrients for crops, provides soil aggregation, increases nutrient exchange, retains moisture, reduces compaction, reduces surface crusting, and increases water infiltration into soil. Components vary in proportion and have many intermediate stages. Plant residues on the soil surface such as leaves, manure, or crop residue are not considered SOM and are usually removed from soil samples by sieving through a 2 mm wire mesh before analysis.

Soil organic matter content can be estimated in the field and tested in a lab to provide estimates for Nitrogen, Phosphorus and Sulfur mineralized and made available for crop/pasture production and to adjust fertilizer recommendations. Soil organic matter impacts the rate of surface applied herbicides along with soil pH necessary to effectively control weeds. Soil organic matter impacts the potential for herbicide carryover for future crops, and amount of lime necessary to raise pH.

Under average conditions in temperate regions (in USA) approximately 1.5 percent of SOM mineralizes yearly for most crops (2% spring planted row crops like corn, 1% small grains such as wheat, and 0.5% perennial grass pasture) while maintaining current organic matter levels on soils with 2 to 5% SOM. Depending on site conditions, management, and climate, mineralization rates and loss of SOM can increase dramatically if temperature, aeration, and moisture conditions are favorable.

Key soil functions SOM provide for include

Nutrient Supply. Upon decomposition, nutrients are released through mineralisation in a plant-available form. While maintaining existing organic matter level, each one percent of SOM in the top 15 cm of a medium textured soil (silt and loam soils with a bulk density of 1.2) releases about 10 – 20 kg of nitrogen, 1 – 2 kg of phosphorus, and 0.4 – 0.8 kg of sulfur per hectare per year.

Another valuable source of nutrient availability beyond normal active organic matter mineralization is from the microbial organic matter pool that occurs when soil dries out and is re-wetted, a significant amount of N-flush can occur from this active fraction over the course of the growing period. This additional N-flush release is measured in the US with a soil CO2 respiration test (Solvita Soil Test). A typical example in a 2% SOM clay/ loam soil with a bulk density of 1.2 g/cm3 has a mineralized organic N of 40 kg/ha/yr. An additional 20 – 30 kgN/ha/yr can be released from microorganisms due to soil wetting and drying.

There is no single explanation for the well known pulse of carbon dioxide or "CO2-Burst" that results as soil re-activates from moisture, but it is associated with a positive nutrient flux. Therefore this application of the Solvita Soil Test represents potential soil N mineralization. The high microbial biomass carbon soil the potential N mineralisation is 75 – 105 kg/ ha over the crop growing season in a humid-temperate environment which may be sufficient N for many crops to grow without added fertiliser. In contrast in soils with low microbial biomass N mineralisation is very low and will not support crop growth.

High soil respiration rates are indicative of high biological activity. This can be a good sign of a healthy soil that readily breaks down organic residues and cycles nutrients needed for crop growth. Solvita® response may go from an inactive condition (0-1) to a very active state (3.5-4.0) as soil respiration increases from desirable management measures such as diverse crop rotations, and no-till.

In some cases, heavily manured soils or soils high in organic content can attain a very high rate. This can be detrimental when decomposition of stable organic matter occurs. It is generally desirable to have at least 3. It typically takes several years for a soil to improve from a low biological status to a more active one. With proper residue management, diverse crop rotations, organic matter additions and avoidance of destructive tillage practices, the time to reach a more optimum condition is shortened.

Water-Holding Capacity. Organic matter behaves somewhat like a sponge. It has the ability to absorb and hold up to 90 percent of its weight in water. Another great advantage of organic matter is that it releases nearly all of the water it holds for use by plants. In contrast, clay holds great quantities of water, but much of it is unavailable to plants.

Soil Aggregation. Organic matter improves soil aggregation, which improves soil structure. With better soil structure, water infiltration through the soil improves, which improves soil's ability to take up and hold water.

Erosion Prevention. Because of increased water infiltration and stable soil aggregates erosion is reduced with increased organic matter.

Estimating organic material needed to increase SOM

The term steady state is where the rate of organic matter addition from crop/pasture residues, roots and manure or other organic materials equals the rate of decomposition. If the rate of organic matter addition is less than the rate of decomposition, SOM will decline and, conversely if the rate of organic matter addition is greater than the rate of decomposition, SOM will increase.

A hectare of soil 20 cm deep weighs approximately 4.8 million kgs, which means that 1 percent SOM weighs about 48,000 kgs per ha. Under average conditions it takes at least 4.2 kg of organic material to decompose into 0.4 kg of organic matter, so it takes at least 91,000 ks (90 tonnes) of organic material applied or returned to the soil to add 1 percent humus or stable organic matter under favorable conditions.

Pasture versus plantation soil organic matter

Soils formed under grass (prairie) vegetation usually have organic matter levels at least twice as high as those formed under forests because organic material is added to topsoil from both top growth and roots that die back every year. Soils formed under forests usually have comparably low organic-matter levels for two main reasons:

1. Trees produce a much smaller root mass per acre than grass plants, and
2. Trees do not die back annually and decompose every year. Instead, much of the organic material in a forest is tied up in the tree's wood rather than being returned to the soil annually.

Soil organic matter generally increases where biomass production is higher and where organic material additions occur. Plant residue with a low C/N ratio (high nitrogen content) decompose more quickly than those with a high C/N ratio and do not increase soil organic matter levels as quickly. Excessive tillage destroys soil aggregates increasing the rate of soil organic matter decomposition. Stable soil aggregates increase active organic matter and protect stable organic matter from rapid microbial decomposition. Measures that increase soil moisture, soil temperature, and optimal aeration accelerate SOM decomposition.

Management measures in a field can either degrade or increase SOM. Some key management measures that can increase SOM are shown below.

- Use of cropping systems that incorporate continuous no-till, cover crops, solid manure or other organic materials, diverse rotations with high residue crops and perennial legumes or grass used in rotation.
- Reducing or eliminating tillage that causes a flush of microbial action that speeds up organic matter decomposition and increases erosion.
- Reduce erosion using appropriate measures. Most SOM is in the topsoil. When soil erodes, organic matter goes with it. Saving soil and SOM go hand in hand.
- Soil-test and fertilize properly. Proper fertilization encourages growth of plants, which increases root and top growth. Increased root growth can help build or maintain SOM, even if you are removing much of the top growth.

- Use of perennial forages provides for annual die back and re-growth of perennial grasses and their extensive root systems and aftermath contributing organic matter to soil. Fibrous root systems of perennial grasses are particularly effective as a binding agent in soil aggregation.

PRACTICES THAT INFLUENCE THE AMOUNT OF ORGANIC MATTER

HUMAN INTERVENTIONS THAT INFLUENCE SOIL ORGANIC MATTER

Various types of human activity decrease soil organic matter contents and biological activity. However, increasing the organic matter content of soils or even maintaining good levels requires a sustained effort that includes returning organic materials to soils and rotations with high-residue crops and deep- or dense-rooting crops. It is especially difficult to raise the organic matter content of soils that are well aerated, such as coarse sands, and soils in warm-hot and arid regions because the added materials decompose rapidly. Soil organic matter levels can be maintained with less organic residue in finetextured soils in cold temperate and moist-wet regions with restricted aeration.

Practices that decrease soil organic matter

Any form of human intervention influences the activity of soil organisms and thus the equilibrium of the system.

Fig. Severe soil erosion removes the potential energy source for soil microbes, resulting in the death of the microbial population and thus of the soil itself.

Management practices that alter the living and nutrient conditions of soil organisms, such as repetitive tillage or burning of vegetation, result in a degradation of their microenvironments. In turn, this results in a reduction of soil biota, both in biomass and diversity. Where there are no longer organisms to decompose soil organic matter and bind soil particles, the soil structure is damaged easily by rain, wind and sun. This can lead to rainwater runoff and soil erosion, removing the potential food for organisms, i.e. the organic matter of the topsoil. Therefore, soil biota are the most important property of the soil, and "when devoid of its biota, the uppermost layer of earth ceases to be soil".

The factors leading to reduction in soil organic matter in an open cycle system can be grouped as factors that result in:

- A decrease in biomass production;
- A decrease in organic matter supply;
- Increased decomposition rates.

DECREASE IN BIOMASS PRODUCTION

Replacement of perennial vegetation

A consequence of clearing forest for agriculture is the disappearance of the litter layer, with a consequent reduction in the numbers and variety of soil organisms. While many temperate forest species appear to adapt well to grassland, the effects of deforestation in the tropics appear to be more marked. Studies have shown that as soil biodiversity declines, adapted species may take over from the indigenous species and the composition may change drastically.

Soil macrofaunal biomass and population density fell to 6 and 17 percent, respectively, in cultivated plots, compared with primary forest in Peruvian Amazonia. In Suriname, the number of animals per square metre has fallen to 36 percent and the diversity of species has fallen to 28 percent compared with primary forest. The indigenous species have largely disappeared, but adapted species have been available for recolonization. The composition of the macrofaunal community has changed drastically.

Replacement of mixed vegetation with monoculture of crops and pastures

The simplification of vegetation and the disappearance of the litter layer under grassland and monocrop production systems lead to a decrease in faunal diversity. Although root systems (especially of grasses) can be extensive and explore vast areas of soil, the root exudates from one single crop will attract only a few different microbial species. This in turn will affect the predator diversity. The more opportunistic pathogen species will be able to acquire space near the crop and cause harm. Continuous cultivation and grazing also leads to compaction of soil layers, which in turn affects the circulation of air. Anaerobic conditions in the soil stimulate the growth of different micro-organisms, resulting in more pathogenic organisms.

High harvest index

One of the consequences of the green revolution was the replacement of indigenous varieties of species with high-yielding varieties (HYVs). These HYVs often produce more grain and less straw, compared with locally developed varieties; the harvest index of the crop (ratio of grain to total plant mass aboveground) is increased. From a production point of view, this is a logical approach. However, this is less desirable from a conservation point of view. Reduced amounts of crop residues remain after harvest for soil cover and organic matter, or for grazing of livestock (which results in manure). Moreover, where animals graze the residues, even less remains for conservation purposes.

Use of bare fallow

Traditionally, a fallow period is used after a period of crop production to give the land some "rest" and to regenerate its original state of productivity. Usually, this is necessary in production systems that have drawn down the nutrient supply and altered the soil biota significantly, such as in slash-and-burn systems or conventional tillage systems.

Some farmers use bare fallow to regenerate their lands. However, apart from spontaneous weed growth, this means there is no energy source for the soil biota present on the land. Instead of recovering the soil food web, the soil organic matter is degraded further and the lack of cover can result in severe erosion and runoff when the rains start after the dry season.

DECREASE IN ORGANIC MATTER SUPPLY

Burning of natural vegetation and crop residues

The burning of maize, rice and other crop residues in the field is a common practice. Residues are usually burned to help control insects or diseases or to make fieldwork easier in the following season. Burning destroys the litter layer and so diminishes the amount of organic matter returned to the soil. The organisms that inhabit the surface soil and litter layer are also eliminated. For future decomposition to take place, energy has to be invested first in rebuilding the microbial community before plant nutrients can be released. Similarly, fallow lands and bush are burned before cultivation. This provides a rapid supply of P to stimulate seed germination. However, the associated loss of nutrients, organic matter and soil biological activity has severe long-term consequences.

Overgrazing

There is a tendency throughout the world to overstock grazing land above its carrying capacity. Cows, draught animals and small ruminants graze on communal grazing areas and on roadsides, stream banks and other public land. Overgrazing destroys the most palatable and useful species in the plant mixture and reduces the density of the plant cover, thereby increasing the erosion hazard and reducing the nutritive value and the carrying capacity of the land.

Removal of crop residues

Many farmers remove residues from the field for use as animal feed and bedding or to make compost. Later, these residues return to contribute to soil fertility as manures or composts. However, residues are sometimes removed from the field and not returned. This removal of plant material impoverishes the soil as it is no longer possible to recycle the plant nutrients present in the residues.

INCREASED DECOMPOSITION RATES

Tillage practices

Tillage is one of the major practices that reduces the organic matter level in the soil. Each time the soil is tilled, it is aerated. As the decomposition of organic matter and the liberation of C are aerobic processes, the oxygen stimulates or speeds up the action of soil microbes, which feed on organic matter.

This means that:

- When ploughed, the residues are incorporated in the soil together with air and come into contact with many micro-organisms, which accelerates the carbon cycle. The decomposition is faster, resulting in the formation of less stable humus and an increased liberation of CO_2 to the atmosphere, and thus a reduction in organic matter.
- The residues on the soil surface slow the carbon cycle because they are exposed to fewer micro-organisms and thus wane more slowly, resulting in the production of humus (which is more stable), and liberating less CO_2 to the atmosphere.

Table. Tillage induced flush of decomposition of organic matter

Type of tillage	Organic matter lost in 19 days (kg/ha)
Mouldboard plough + disc harrow (2x)	4 300
Mouldboard plough	2 230
Disc harrow	1 840
Chisel plough	1 720
Direct seeding	860

In terms of short-term organic matter loss, the more a soil is tilled, the more the organic matter is broken down. There are also longer-term losses, attributed to repeated, annual cultivation. Cropping systems that return little residue to the soil accelerate this decline. Many modern cropping systems combine frequent tillage with small amounts of residue, with resultant reductions in the organic matter content of many soils. Historically, manure application (from farm livestock) was common, and it was a dynamic way of maintaining organic matter levels despite repeated cultivation and low residue returns to the soil. Increased on-farm mechanization has reduced livestock numbers, so this source of organic material has been reduced considerably.

Organic matter production and conservation is affected dramatically by conventional tillage, which not only decreases soil organic matter but also increases the potential for erosion by wind and water. The impact occurs in many ways:

- Ploughing leaves no residues on the soil surface to lessen the impact of rain.
- Ploughing reduces the quantity of food sources for earthworms and disturbs their burrows and living space, hence populations of certain species decrease drastically. Moreover, reduction of earthworm numbers reduces their impact, through burrowing, in increasing porosity and aeration (particularly continuous macropores) and lowers their ability to bury and incorporate plant residues, which facilitates rapid decomposition of organic matter.
- Tillage by repeated hoeing or discing smoothes the surface and destroys natural soil aggregates and channels that connect the surface with the subsoil, leaving the soil susceptible to erosion. Old root channels and earthworm holes are eliminated, as are the cracks between natural aggregates. The large pores, the ones destroyed by conventional tillage practices, are necessary to conduct water into the soil during rainfall.
- The development of a plough pan or hoe pan, a layer of compacted soil resulting from smearing action at the bottom of the plough or hoe, may retard both root penetration and water infiltration.
- Ploughing or discing under dry conditions exacerbates the pulverization of the soil, causing the soil surface to crust more easily, leading to greater water runoff and erosion. This is exacerbated by reduced soil surface roughness, which leaves few depressions for temporary storage of water during intense storms.
- Increased runoff during rainstorms may also increase the possibility of drought stress later in the season, because water that runs off the field does not infiltrate into the soil to remain available to plants.

In some circumstances, imbalances of certain soil organisms can disrupt soil structure and processes, e.g. certain earthworm species in rice fields or pastures.

Drainage

Decomposition of organic matter occurs more slowly in poorly aerated soils, where oxygen is limiting or absent, compared with well-aerated soils. For this reason, organic matter accumulates in wet soil environments. Soil drainage is determined strongly by topography - soils in depressions at the bottom of hills tend to remain wet for extended periods of time because they receive water (and sediments) from upslope. Soils may also have a layer in the subsoil that inhibits drainage, again exacerbating waterlogging and reduction in organic matter decomposition. In a permanently waterlogged soil, one of the major structural parts of plants, lignin, does not decompose at all. The ultimate consequence of extremely wet or swampy conditions is the development of organic (peat or muck) soils, with organic matter contents of more than 30 percent. Where soils are drained artificially for agricultural or other uses, the soil organic matter decomposes rapidly.

Fertilizer and pesticide use

Initially, the use of fertilizer and pesticides enhances crop development and thus production of biomass (especially important on depleted soils). However, the use of some fertilizers, especially N fertilizers, and pesticides can boost micro-organism activity and thus decomposition of organic matter. The chemicals provide the microorganisms with easy-to-use N components. This is especially important where the C: N ratio of the soil organic matter is high and thus decomposition is slowed by a lack of N.

PRACTICES THAT INCREASE SOIL ORGANIC MATTER

Increased concern about the environmental and economic impacts of conventional crop production has stimulated interest in alternative systems. Central to such systems is the need to promote and maintain soil biological processes and minimize fossil fuel inputs in the form of fertilizers, pesticides and mechanical cultivation. All activities aimed at the increase of organic matter in the soil help in creating a new equilibrium in the agro-ecosystem.

For a system of natural resource management to be balanced, and thus sustainable, it must be able to withstand sharp climatic fluctuations, and to evolve steadily in response to social changes and changes in the costs and availability of inputs of land, labour and knowledge. The more diverse and complex an agricultural system is, the more stable and sustainable it will be in the face of unpredictable vagaries of climate and market. Thus, annual crops, woody perennials and nonwoody perennials may be combined in various ways with livestock or trees, or both, in what are now commonly called agrosilvipastoral systems.

Different approaches are required for different soil and climate conditions. However, the activities will be based on the same principle:

increasing biomass production in order to build active organic matter. Active organic matter provides habitat and food for beneficial soil organisms that help build soil structure and porosity, provide nutrients to plants, and improve the water holding capacity of the soil.

Several cases have demonstrated that it is possible to restore organic matter levels in the soil. Activities that promote the accumulation and supply of organic matter, such as the use of cover crops and refraining from burning, and those that reduce decomposition rates, such as reduced and zero tillage, lead to an increase in the organic matter content in the soil.

Fig. Evaluation of the organic matter content of a soil in Paraná

INCREASED BIOMASS PRODUCTION

Increased water availability for plants: water harvesting and irrigation

In dry conditions, water may be provided through irrigation or water harvesting. The increased water availability enhances biomass production, soil biological activity and plant residues and roots that provide organic matter.

The concept of water harvesting includes various technologies for runoff management and utilization. It involves capture of runoff (in some cases through treating the upstream capture area), and its concentration on a runon area for use by a specific crop (annual or perennial) in order to enhance crop growth and yields, or its collection and storage for supplementary irrigation or domestic or livestock purposes. The objective of designing a water harvesting system is to obtain the best ratio of the area yielding runoff to either the area where runoff is being directed or the capacity of the storage structure (volume of water collected). In this way, the water captured for crop production during runoff periods can be stored either directly in the soil for subsequent use by plants or in small farm reservoirs or collection tanks. This aids stabilization of crop production by enhancing soil moisture availability or allowing irrigation during a dry period within the rainy season or by extending crop production into the dry season. Some factors to be considered regarding these runoff farming systems and reservoirs include: site selection, watershed size and condition, rainfall distribution and runoff, and water requirements of crops. Where a minimum water depth of about 1 m can be maintained in a reservoir, fish can be raised to provide additional food.

Numerous water harvesting systems have been developed over the centuries, especially in arid areas. The principle of collecting runoff for crop production is also inherent to many other soil and water conservation technologies that apply the concept of runoff and runon areas at a microwatershed level, such as negarims, trapezoidal or "eyebrow" bunds and tied ridges.

Balanced fertilization

Where the supply of nutrients in the soil is ample, crops are more likely to grow well and produce large amounts of biomass. Fertilizers are needed in those cases where nutrients in the soil are lacking and cannot produce healthy crops and sufficient biomass. Most soils in sub-Saharan Africa (SSA) are deficient in P. P is required not only for plant growth but also for N fixation. Unbalanced fertilization, for example mainly with N, may result in more weed competition, higher pest incidence and loss of quality of the product. Unbalanced fertilization eventually leads to unhealthy plants. Therefore, fertilizers should be applied in sufficient quantities and in balanced proportions. The efficiency of fertilizer use will be high where the organic matter content of the soil is also high. In very poor or depleted soils, crops use fertilizer applications inefficiently. When soil organic matter levels are restored, fertilizer can help maintain the revolving fund of nutrients in the soil by increasing crop yields and, consequently, the amount of residues returned to the soil.

Cover crops

Growing cover crops is one of the best practices for improving organic matter levels and, hence, soil quality. The benefits of growing cover crops include:

- They prevent erosion by anchoring soil and lessening the impact of raindrops.
- They add plant material to the soil for organic matter replenishment.
- Some, e.g. rye, bind excess nutrients in the soil and prevent leaching.
- Some, especially leguminous species, e.g. hairy vetch, fix N in the soil for future use.
- Most provide habitat for beneficial insects and other organisms.
- They moderate soil temperatures and, hence, protect soil organisms.

A range of crops can be used as vegetative cover, e.g. grains, legumes and oil crops. All have the potential to provide great benefit to the soil. However, some crops emphasize certain benefits; a useful consideration when planning a rotation scheme. It is important to start the first years with (cover) crops that cover the surface with a large amount of residues that decompose slowly (because of the high C:N ratio). Grasses and cereals are most appropriate for this stage, also because of their intensive rooting system, which improves the soil structure rapidly.

In the following years, when soil health has begun to improve, legumes can be incorporated in the rotation. Leguminous crops enrich the soil with N and their residues decompose rapidly because of their low C:N ratio. Later, when the system is stabilized, it is possible to include cover crops with an economic function, e.g. livestock fodder.

The selection of cover crops should depend on the presence of high levels of lignin and phenolic acids. These give the residues a higher resistance to decomposition and thus result in soil protection for a longer period and the production of more stable

Another determining factor in the dynamics of residue composition is the biochemical composition of the residues. Depending on species, their chemical components and the time and way of managing them, there will be differences in decomposition rates. The grain species (oats and wheat) show more resistance than common vetch (legume) to decomposition. The latter has a lower C:N ratio and a lower lignin content and is thus subject to a rapid decomposition.

Agricultural production systems in which residues are left on the soil surface, such as direct seeding and the use of cover crops, stimulate the development and activity of soil fauna at many levels.

The term green manure is often used to indicate the same plant species that are used as cover crops. However, green manure refers specifically to a crop in the rotation grown for incorporation of the non-decomposed vegetative matter in the soil. While this practice is used specifically to add organic matter, this is not the most effective use of organic matter (especially in hot climates) for two reasons:

- Mechanical disturbance of the soil should be avoided as much as possible.
- When biomass is incorporated in the soil all at one time, there is a short period of high microbial activity in decomposing the material. This results in the sudden release of a large quantity of nutrients that cannot be captured by the seedlings of the following crop and is thus lost from the system.

In general, the greater the production of green manure or crop biomass, the greater is the microbial, mesofauna and macrofauna population of the soil - from fungi and micro-organisms to earthworms and termites. The dynamics of surface residue decomposition depend *inter alia* on the activity of micro-organisms and also on soil mesofauna and macrofauna. The macrofauna consists mainly of earthworms, beetles, termites, ants, millipedes, spiders, snails and slugs. These organisms help integrate the residues into the soil and improve soil structure, porosity, water infiltration, and through-flow through the creation of burrows, ingestion and secretions.

The natural incorporation of cover-crop and weed residues from the soil surface to deeper layers in the soil by soil macrofauna is a slow process. The activity of microorganisms is regulated by the activity of the macrofauna, because the latter provide them with food and air through their bioturbation activities. In this way, nutrients are released slowly and can provide the crop with nutrients over a longer period. At the same time, the soil is covered for a long time by the residues and is protected against the impact of rain and sun.

Improved vegetative stands

In many places, low plant densities limit crop yields. Wide plant spacing is often practised as "a way to return power to the soil" or "to give the soil some rest", but in reality it is an indicator that the soil is impoverished. Plant spacing is usually determined by farmers in relation to soil fertility and available water or expected rainfall (unless standard recommendations are enforced by extension).

This means that plants are often spaced widely on depleted soils in arid and semi-arid regions with a view to ensuring an adequate provision of plant nutrients and water for all plants.

However, it is important to maintain the recommended plant spacing in order to optimize biomass production and rooting density and, hence, organic matter for food, moisture retention and habitat for soil organisms. Once the crop is established, reduced sunlight between closer crop rows may also reduce regrowth of weeds.

Planting pits are a way of increasing biomass production and crop yields on severely degraded land in semi-arid conditions. Rainfall is concentrated near the plants, and soil faunal activity and organic matter accumulation are concentrated in the planting pits. Planting pits have been introduced successfully in Zambia as a conservation practice for smallholder farmers, who do not have fertilizers or tractor services available to them.

Agroforestry and alley cropping

Agroforestry is a collective name for land-use systems where woody perennials (trees, shrubs, palms, etc.) are integrated in the farming system. Alley cropping is an agroforestry system in which crops are grown between rows of planted woody shrubs or trees. These are pruned during the cropping season to provide green manure and to minimize shading of crops.

Agroforestry covers a wide range of systems combining food crops, forestry and pasture species in different ways (agrosilviculture, silvipasture, agrosilvipasture and multipurpose forest production). There are two different approaches to agroforestry. One uses agricultural crops or pasture as a transitional means of utilizing the land until forest plantations are fully established. The other is to integrate trees and shrubs permanently into the crop or animal production system, to the benefit of both crop production and land resource protection. Thus, agroforestry encompasses many traditional land-use systems such as home gardens, shifting cultivation and bush fallow systems.

Alley cropping can be considered an improved bush fallow system. Small trees or shrubs are planted in cropland in rows, preferably along the contour (even where east-west orientation of the rows may minimize shading of crops). The optimal spacing between rows depends on: slope; soil type and its susceptibility to erosion; rainfall; crop species; and the soil and crop management system.

Besides adding organic matter to the system, perennial trees and shrubs recycle plant nutrients from deeper soil layers through their rooting system. Through litter and pruning, these can be used again by annual crops. Probably the most important contribution of perennials in a production system lies in the fact that throughout the whole year their roots excrete root exudates and decaying root cells, which in turn are used as an energy source by soil microorganisms. The food web in the soil is maintained, even during dry seasons when no annual crops are grown. The result is that soil biota are in place to provide the crop with nutrients at the beginning of the next cropping season.

Direct seeding is the easiest and cheapest way of establishing hedgerows around fields or in the fields (alleys). However, emerging seedlings may not be able to compete with weeds without additional care. Therefore, starting plant growth in a nursery and transplanting may be necessary for some species. Other species may be established by cuttings. With good establishment, the plants will be better able to withstand both dry spells and browsing by livestock. Crucial to a successful establishment of the hedgerow is that the selected plants should be tall enough to outgrow the weeds at the time of the first crop harvest.

During the cropping season, hedgerow pruning is needed in order to avoid shading of the crop. The timing, frequency and extent of pruning depend on the species used and the season. As a general rule, the lower the hedgerows and the taller the crop, the less frequently is pruning required. Fastgrowing plants such as *Leucaena leucocephala* and *Gliricidia sepium* may require pruning every six weeks during the cropping season. They are often pruned to a height of about 50 cm. Care must be exercised as too frequent pruning can result in tree dieback.

The integration of trees and woody shrubs into the cropping system offers additional uses and many benefits, as mentioned by farmers using the Quezungual system in Honduras. However, farmers with short-term land tenure may not be interested in these benefits. Furthermore, the plantation of trees sometimes has an effect on the land tenure status; therefore tenants may not be allowed to establish trees on agricultural land. Agroforestry systems can also inhibit mechanization and may need increased labour inputs, especially for hedgerow pruning.

The hedgerow species have to be selected carefully in order to avoid negative impacts on crop production because of the complex relationships (competition for light, water and nutrients, allelopathy, occurrence of pest and diseases, etc.) that are inherent to agroforestry systems. Many farmers may consider the hedgerows as not useful, especially where their positive effects are not secure or visible. Where livestock are allowed to graze freely, it can be difficult to establish hedgerows without taking special measures to protect the young plants. Fencing or control of grazing animals may require collective efforts and agreement by the local community.

The increased labour requirement, the reduced cropped area and the difficulty of mechanization may make alley cropping uneconomic unless the hedgerow species produce direct benefits such as fruits, fuelwood or poles/ timber for construction purposes (in addition to the nutrient recycling and erosion control effects).

Reforestation and afforestation

Afforestation means the establishment of a forest on land that has not grown trees recently. It can serve two principal soil and water conservation purposes: protection of erosion-prone areas, and revegetation and rehabilitation of degraded land. Afforestation is specifically used to provide protective cover in vulnerable, steep and mountainous areas. Afforestation helps to replenish timber resources and provide fuelwood and fodder.

The establishment of a forest cover under good management is an effective means of increasing organic matter production. However, the land must have the productive capacity to support an appropriate forest type, which differs according to climate, soil, slope and the specific purpose of the forest (timber production, livestock grazing, etc.). Therefore, the choice of species and the selection of an appropriate site are of particular importance for successful afforestation.

The procurement of adequate quantities of good quality seed of the species and provenances (adapted varieties) required is a prerequisite for any afforestation effort. However, it is often difficult to find suitable and reliable sources of such seeds. A number of species require special pre-treatment of the seed or seedling in order to achieve satisfactory germination and uniform stand. Such treatment may consist of soaking the seed in water for varying lengths of time, alternate soaking and drying, scarifying or chipping the seed coat to render it permeable to water, plunging the seed into boiling water or even boiling it for a short time. Some tree seedlings may need a mycorrhizal treatment when planted in soils that are deprived of associated mycorrhizae species as well as rhizobium species (e.g. *Casuarina* in Senegalese sandy soils). The aim is to ensure that good numbers of plants germinate and that germination after sowing is both rapid and uniform.

Afforestation can be achieved by direct sowing or replanting young plants from a nursery. The main advantage of direct sowing is the reduced cost. However, this is usually much less reliable and is only justified where:

- Seed is plentiful and cheap;
- Adequate germination under field conditions can be relied on;
- The seedlings send down a deep tap root rapidly and are able to withstand adverse climatic conditions in the time after germination;
- The rate of growth is sufficiently fast to make a prolonged period of tending and weeding unnecessary.

Regeneration of natural vegetation

Regeneration of natural grasslands and forest areas increases biomass production and improves the plant species diversity, resulting in more diverse soil biota and other associated beneficial organisms. Natural regeneration may be more reliable where land is not very productive. In some cases, natural regeneration of a given area may lead to the infestation of plots by weeds. Increasingly, natural vegetation is being recognized for its multipurpose benefits, for example, fuelwood, fibre, biocontrol (e.g. neem) and medicinal species, as well as restoration of soil fertility (*Acacia albida* and other leguminous species) and habitats for various beneficial species (pollinators and natural enemies) as well as wildlife.

INCREASED ORGANIC MATTER SUPPLY

Protection from fire

Burning affects organic matter recycling significantly. Fire destroys almost all organic materials on the land surface except for tree trunks and large branches. In addition, the surface soil is sterilized, loses part of its organic matter, the population of soil microfauna and macrofauna is reduced, and no ready-to-use organic matter is available for rapid restoration of the populations. However, this practice is widely used (e.g. in Africa) in order to enhance pasture regrowth for livestock (using residual P), to control pests and diseases, and even to catch small animals for food. A specific and difficult case is the burning of sugar cane before harvesting. It has both a technical dimension (CO_2 and greenhouse gas emissions, mechanization of harvest, sugar content, etc.) and a social dimension (manual cutting, source of survival resources for poor/landless workers).

The damage depends on fire intensity, which is a function of vegetation type and climate conditions and frequency. The costs and benefits of burning and the methods to minimize harmful effects need to be identified with local populations.

Crop residue management

In systems where crop residues are managed well, they:

- Add soil organic matter, which improves the quality of the seedbed and increases the water infiltration and retention capacity of the soil, buffers the pH and facilitates the availability of nutrients;
- Sequester (store) C in the soil;
- Provide nutrients for soil biological activity and plant uptake;
- Capture the rainfall on the surface and thus increase infiltration and the soil moisture content;
- Provide a cover to protect the soil from being eroded;
- Reduce evaporation and avoid desiccation from the soil surface.

Depending on the nature of the following crop, decisions are made as to whether the residues should be distributed evenly over the field or left intact, e.g. where climbing cover crops (e.g. mucuna) use the maize stalks as a trellis.

An even distribution of residues: (i) provides homogenous temperature and humidity conditions at sowing time; (ii) facilitates even sowing, germination and emergence; (iii) minimizes the development of pests and diseases; and (iv) reduces the emergence of weeds through allelopathic effects.

The most appropriate method for managing crop residues depends on the purpose of the crop residues and the experience and equipment available to the farmer. Where the aim is to maintain a mulch over the soil for as long as possible, the biomass is best managed using a knife roller, chain or sledge in order to break it down but not kill it. Where the decomposition process should commence immediately in order to release nutrients, the residues should be slashed or mown and some N applied because dry residues have a high C:N ratio. However, in order to avoid nitrate emission, urea should not be broadcast on the surface but injected where possible.

Utilizing forage by grazing rather than by harvesting

In many places, there is competition for the use of crop residues that can be used as fodder, for roofing, artisan handicrafts, etc. Where residues are to be used for animal feed, either the animals graze the residues directly, or they are stall- or kraal-fed.

Removal of the residues from the field can lead to a considerable loss of organic matter where animal manure is not returned to the field. By controlled grazing, the animal manure is returned in the field without a high labour input.

The experience of Guaymango, El Salvador, demonstrates that it is possible to achieve successful integration of crop and livestock components without creating competition in the allocation of crop residues. The amount of residues produced by the system is enough to serve both as soil cover and as fodder for livestock, mainly because of the use of local sorghum varieties (instead of HYVs) that have a high straw/grain ratio (Choto, Saín and Montenegro, 1995). As farmers value crop residues as soil cover, a fodder market has developed where grazing rights, number of cattle and duration of grazing are traded.

In the northern zone of the United Republic of Tanzania, farmers have found a compromise between using the residues for grazing or soil cover, albeit one that is rather labour intensive. They separate the palatable and non-palatable parts of the crop residues. They use the non-palatable parts to cover the soil and act as food for soil organisms, while they feed the palatable parts to cattle and goats that are kept close to the homestead.

Integrated pest management

As with balanced fertilization, proper pest and disease management results in healthy crops. Healthy crops produce optimal biomass, which is necessary for organic matter production in the soil. Diversified cropping and mixed crop-livestock systems enhance biological control of pests and diseases through species interactions.

Fig. Farmer giving fodder to cattle

Through integrated production and pest management farmers learn how to maintain a healthy environment for their crops. They learn to examine their crops regularly in order to observe ratios of pests to natural enemies (beneficial predators) and cases of damage, and on that basis to make decisions as to whether it is necessary to use natural treatments (using local products such as neem or tobacco) or chemical treatments and the required applications.

Applying animal manure or other carbon-rich wastes

Any application of animal manure, slurry or other carbon-rich wastes, such as coffee-berry pulp, improves the organic matter content of the soil. In some cases, it is better to allow a period of decomposition before application to the field. Any addition of carbon-rich compounds immobilizes available N in the soil temporarily, as micro-organisms need both C and N for their growth and development. Animal manure is usually rich in N, so N immobilization is minimal. Where straw makes up part of the manure, a decomposition period avoids N immobilization in the field.

Compost

Composting is a technology for recycling organic materials in order to achieve enhanced agricultural production. Biological and chemical processes accelerate the rate of decomposition and transform organic materials into a more stable humus form for application to the soil. Composting proceeds under controlled conditions in compost heaps and pits.

Compost heaps should have a minimum size of 1 m^3 and are suitable for more humid environments where there is potential for watering the compost. Compost pits should be no deeper than 70 cm and should be underlain with rough material for good aeration of the compost. Pits are suitable for drier environments where the compost may desiccate. Dry composting relies on covering the compost with soil and creating an anaerobic environment. However, this is a slower process than the more usual moist aerobic process. The ratio of C to N in the compost pile is important for optimizing microbial activity. Thus, a mixture of soft, green and brown, tougher material is used. Ash and phosphate rock are often added to accelerate the process.

Composting can complement certain crop rotations and agroforestry systems. It can be used efficiently in planting pits and nurseries. It is very similar in composition to soil organic matter. It breaks down slowly in the soil and is very good at improving the physical condition of the soil (whereas manure and sludge may break down fairly quickly, releasing a flush of nutrients for plant growth). In many circumstances, it takes time to rejuvenate a poor soil using these practices because the amount of organic material being added is small relative to the mineral proportion of the soil.

Successful composting depends upon the sufficient availability of organic materials, water, manure and "cheap" labour. Where these inputs are guaranteed, composting can be an important method of sustainable and productive agriculture. It has ameliorative effects on soil fertility and physical, chemical and biological soil properties. Well-made compost contains all the nutrients needed by plants. It can be used to maintain and improve soil fertility as well as to regenerate degraded soil. However, materials for compost production may be in short supply and the technology demands high labour inputs for proper compost production and application. Therefore, compost application may be restricted to certain crops and limited application areas, e.g. vegetable production in home gardens.

Mulch or permanent soil cover

One way to improve the condition of the soil is to mulch the area requiring amelioration. Mulches are materials placed on the soil surface to protect it against raindrop impact and erosion, and to enhance its fertility. Crop residue mulching is a system of maintaining a protective cover of vegetative residues such as straw, maize stalks, palm fronds and stubble on the soil surface.

The system is particularly valuable where a satisfactory plant cover cannot be established rapidly when erosion risk is greatest.

Mulching adds organic matter to the soil, reduces weed growth, and virtually eliminates erosion during the period when the ground is covered with mulch.

There are two principal mulching systems:

- *In situ* mulching systems - plant residues remain where they fall on the ground ;
- Cut-and-carry mulching systems - plant residues are brought from elsewhere and used as mulch.

Crop residue mulching has numerous positive effects on crop production. However, it may require a change in existing cropping practices. For example, farmers may conventionally burn crop residues instead of returning them to the soil. *In situ* mulching depends on the design of appropriate cropping systems and crop rotations, which have to be integrated with the farming system. The greater labour demands of cut-and-carry systems represent a major constraint. Mulch may be more relevant in home gardens or for valuable horticulture crops than in less intensive farming systems.

Mulch affects the soil life. Holland and Coleman (1987) have demonstrated that litter placement on the soil surface (as opposed to incorporation with ploughing) increased the ratio of fungi to bacteria - the reason being that fungi have a higher carbon assimilation efficiency than bacteria. In addition, it encourages bioturbating (mixing) effects of macrofauna that pull the materials into surface layers of the soil.

DECREASED DECOMPOSITION RATES

Reduced or zero tillage

Repetitive tillage degrades the soil structure and its potential to hold moisture, reduces the amount of organic matter in the soil, breaks up aggregates, and reduces the population of soil fauna such as earthworms that contribute to nutrient cycling and soil structure.

Avoiding mechanical soil disturbance implies growing crops without mechanical seedbed preparation or soil disturbance since the harvest of the previous crop. The term zero tillage is used for this practice synonymously with terms such as no-till farming, no tillage, direct drilling, and direct seeding.

Compared with conventional tillage, reduced or zero tillage has two advantages with respect to soil organic matter. Conventional tillage stimulates the heterotrophic microbiological activity through soil aeration, resulting in increased mineralization rate. Through breakdown of soil structure, it decreases upward and downward movements of soil fauna, such as earthworms, which are largely responsible for "humus" production through the ingestion of fresh residues. Reduced or zero tillage regulates heterotrophic microbiological activity because the pore atmosphere is richer in CO_2/O_2, and facilitates the activity of the "humifiers".

Tillage has become the most common method to control weeds. However, mulching is a more environmentally sound practice than tillage for weed control. The loose soil that results from tilling has less structure than before; the appearance is deceptive. Subsequent traffic or heavy rain soon packs this loosened soil, not only negating the expensive cultivation that produced the loose soil but also culminating in a degraded environment for water entry, seed germination and root growth. Further cultivation is then required to re-loosen the soil; more expense with the same outcome - subsequent repacking and degraded soil structure. This is a typical "downward spiral" of conventional agriculture. Moreover, tillage when the soil is too moist or too dry leads to compaction or pulverization of soil; but farmers may not have the option to wait for optimal conditions.

Severe, accelerated soil erosion and the high costs in terms of labour and energy associated with plough-based methods of seedbed preparation have led to the widespread adoption of no- or zero-tillage systems for cropping in temperate and tropical climates. In no-tillage systems, the crop is sown into a soil left undisturbed since the harvest of the previous crop. Crop residue mulch is maintained and anchored firmly to the ground. Weed control relies on mechanical slashing or cover crops. Contact herbicides are also used in some cases. In reduced- or zero-tillage systems, soil fauna resume their bioturbating activities gradually. These loosen the soil and mix the soil components (also known as biotillage). The additional benefit of the increased soil organic matter and burrowing is the creation of a stable and porous soil structure without expensive, time-consuming and potentially degrading cultivations.

In zero-tillage systems, the action of soil macrofauna gradually incorporate cover crop and weed residues from the soil surface down into the soil. The activity of microorganisms is also regulated by the activity of the macrofauna, which provide them with food and air through their burrows. In this way, nutrients are released slowly and can provide the following crop with nutrients.

Several authors have demonstrated that some crop rotations and zero tillage favour *Bradyrhizobia* populations, nodulation and thus N fixation and yield. A 200-300-percent increase in population size of root nodule bacteria in a zero-tillage system compared with conventional tillage. The presence of soybean in the crop rotation resulted in a fivefold to tenfold increase in population size of the same bacteria compared with cropping systems without soybean.

Strictly speaking, the term zero tillage applies to methods involving no soil disturbance whatsoever, a condition that may be difficult to achieve. Broadcasting of seed is one way of applying zero tillage. The seed is broadcast over the previous crop residues and, where necessary, the residues are shaken to ensure that the seed falls on the soil surface.

In direct drilling, seeds such as maize, sorghum, soybean, wheat and barley are sown directly into shallow furrows cut into the previous crop residues. Weeds are controlled mechanically with a knife, which knocks down the plants and breaks their stems, or chemically with herbicides.

Traditional practices such as the burning of crop residues may inhibit the introduction of no-tillage systems. In many situations, a conflict exists between leaving crop residues on the surface or feeding them to livestock in the dry season when there is a shortage of fodder.

Mechanical soil disturbance also includes soil compaction through wheel impact of machinery, especially important in large-scale mechanized agriculture, e.g. plantations or biannual crops (cotton). In a zero-tillage farming system, consideration must be given to reducing both the random placement of tyres/wheels in fields as well as the potential for compaction from animal hooves. Pietola, Horn and Yli-Halla (2003) reported the destructive effect of cattle trampling on the soil structure. Proffitt, Bendotti and McGarry (1995) demonstrated the almost total loss of soil porosity in the soil surface as a result of trampling by sheep. There is a belief that draught animals cause less land degradation than tractors. However, there are reports of soil compaction on smallholder farming enterprises in both Malawi and Bangladesh. The hooves of draught animals and the shearing effect of ploughs or hand hoes, which are used repeatedly at a constant depth, can cause severe compacted layers. Grazing animals should be removed from zero-till fields in moist-wet soil conditions as the compaction risk is greatest at these times.

Fig. "Frijol tapado" or broadcast beans on the residues of maize; a common practice in Latin America.

Fig. Maize seedling directly drilled in residues of wheat.

Controlled traffic, where the wheels of all in-field equipment follow permanent, defined tracks, ensures that compaction is restricted to specific known areas. Alternatively, flotation tyres (low ground-pressure tyres) should be fitted to all large tractors, harvesters, in-field grain bins, etc. in order to reduce their compacting potential.

Recent research has demonstrated the devastating effects of compaction from wheel impact on the occurrence and survival of eartworms. Earthworm incidence was greater under controlled traffic than under wheeled traffic. The immediate effects of wheeling and tillage on the earthworm population. It appears that wheeling has the most detrimental effect on earthworm survival and that where wheeling is followed by tillage the survival rate is much greater. This may indicate that earthworms are able to survive an initial compaction in the field as long as it is relieved immediately. Where it is not, earthworms are inmobilized and unable to find air and nutrients.

10

Management of Organic Wastes and Green Manures

MANURE MANAGEMENT AND COMPOSTING

Manure is a valuable resource on an organic farm. Livestock are inefficient in extracting nutrients from feedstuffs; typically, 75-90 per cent of major nutrients that are fed to livestock pass directly through the animal into the manure. The extent to which these nutrients can be returned to the soil and made available to subsequent crops will depend on the way the manure is stored and handled.

Organic farmers rarely apply raw manure to their fields; they use composted manure. The composting process imitates the decomposition of organic matter on the surface layer of the soil to turn raw manure into humus. Composted manure slowly releases its nutrients into the soil, enhancing the soil microbiological life, whereas the highly-soluble nutrients in raw manure are quickly leached away and can damage both the soil biology and the crop.

Standards for organic production allow the use of manure that has been stacked and aged for at least six months unless it has been brought in from off the farm, in which case composting is required. Raw manure can also be used but only on perennials or crops not for human consumption. Raw and aged manure should be spread when the soil is warm enough (i.e., when plants are growing) for the micro-organisms to be active.

STORAGE OF SOLID MANURE

As composted manure is the primary source of fertilizer for an organic farm, care should be taken that nutrients are not lost from the raw manure. Losses of nitrogen can occur as soon as the urine hits the concrete under the animal. Therefore use generous amounts of bedding to soak up liquid wastes and to provide a carbon source for composting the manure. Considerable losses, especially of potassium, can occur as a result of leaching and runoff during storage. This is not only a serious pollution problem, but also a waste of valuable nutrients.

Plans and information regarding manure storage and handling are available from your provincial department of agriculture.

Solid manure storage systems should contain three parts:

1. An enclosed area, on a sloped cement pad, preferably under a cover;
2. A perimeter curb to contain and direct the liquids to the low end; and
3. A pit to contain the liquid runoff until it can be pumped and spread on the land.

Locate the manure storage in such a way as to allow for future expansion of the livestock operation. Also, do not block livestock or vehicle flow. Don't locate the storage within 30 metres (100 feet) of a well or excavate into ground with a high water table. Adjacent existing tile drains should be blocked off. Check your local zoning by-law to ensure that you comply with any municipal requirements. The*Agricultural Code of Practice* contains information on siting manure storages. Minimum separation distances are provided for distances to neighbors' houses, lot lines and roads.

COMPOSTING

Composting is the process of decomposing organic matter, whether manure, crop residue or municipal wastes, by a mixed microbial population in a warm, moist aerobic environment. The organic matter is decomposed by the successive action of bacteria, fungi and actinomycetes. In the final stages of decomposition, redworms (or manure worms) assist in the production of stable humus which is the final stage of the composting process.

Young compost that has not reached the stable humus stage, will be high in effective humus and available nutrients but low in stable, colloidal humus. Mature compost, which is close to the final decomposition stage, will have a higher proportion of stable humus, and will be considerably reduced in bulk. Compost at various stages, from young to fully-aged, may be used according to the needs of the soil and the crop.

The nutritive and other benefits of the material will depend very much on the source materials, the conditions under which it was made and the maturity of the compost when it is applied. Young or medium compost will encourage biological activity in the soil. Mature compost will make a greater contribution to soil organic matter levels and soil structure. In general, however, the process results in a net improvement to soil fertility, compared to an application of manure. For example, a field application of 30 tonnes of farmyard manure might supply 3 tonnes of stable humus after a 4-5 year breakdown period. The same material applied as about 22 tonnes of young to medium compost supplies about 4 tonnes, or applied as 15 tonnes of mature compost supplies 5 tonnes of stable humus after the same period.

ADVANTAGES OF COMPOSTING

The additional storage and handling requirements involved in the production of compost are offset by the advantages of compost to the organic farmer. These advantages are as follows:

1. Compost supports and encourages the growth of earthworms, bacteria, fungi and other micro-organisms and adds organic matter to the soil. In this way, compost improves the biological, physical and chemical properties of the soil. In comparison, raw manure also adds organic matter but can cause a period of disruption to the soil life by creating an imbalance of nutrients.
2. Manure is acidic; composting increases the pH of the material which can help make the soil a better environment for plant growth.
3. The composting process stabilizes the volatile nitrogen of raw manure into large protein particles and thereby reduces losses.
4. Compost returns nitrogen, phosphorous, potassium, calcium, magnesium and the micronutrients back to the soil. Amounts vary, but a well-prepared mature compost may contain 7.5-15 kg/t (15-30 lbs/ton) N, 2.5-5 kg/t (5-10 lbs/ton) P2O5 and 15kg/t (30 lbs/ton) K2O.
5. The nutrients from mature compost are released to the plants slowly and steadily. The benefits will last for more than one season.
6. The nature of the material and the fungal/actinomycete mycelia contained in the compost and stimulated in the soil by its application help to bind the soil particles into crumbs, greatly increasing the stability of the soil to wind and water erosion.
7. Compost has a lower density, 400-600 kg/m3 compared with typical manure that may be 400-1000 kg/m3. Handling is easier and fewer trips are made to the field.
8. Odor is reduced.
9. Weed seeds are reduced by a combination of factors including the heat of the compost pile, rotting and premature germination. (Any weeds found growing on the pile should be destroyed before they go to seed.)
10. Fly eggs are killed and plant and animal pathogens are reduced if the high heat method of composting is used to raise the temperature of the pile to 60°C.
11. Raw manure is one of the primary culprits for pollution of the waterways, and odor from farms is considered an increasing problem in the rural areas. Composting raw manure reduces these problems.

METHODS

Making good compost depends on having the proper sources of nutrients with a balance of carbon and nitrogen, keeping the pile of compost moist and making sure that there is adequate aeration. The compost pile can heat up to 60-70°C due to the microbial activity. However, high temperatures will result in substantial losses of nitrogen in the form of ammonia gas. Farmers with many years experience at compost-making recommend that temperatures are kept below 50°C to avoid overheating and nutrient losses.

MATERIALS TO COMPOST

The most commonly used materials for the compost pile are manure mixed with livestock bedding. When the bedding (which is predominantly carbon) is mixed with the raw manure (which is an excellent source of nitrogen), you achieve the balance of carbon to nitrogen needed to begin the composting process.

Bedding materials vary in their carbon:nitrogen (C:N) ratio from about 80:1 in straw to 200:1 or more in sawdust or shavings. Bedding with a high content of wasted hay, typical for sheep pens, will have a lower C:N ratio. If the bedding:manure ratio is high, and the manure is very dry as with horse operations, it might be beneficial to water the material with a high N additive such as liquid manure imported from a hog operation. In practice, this is difficult to do.

Provided that it contains no hazardous substances and the correct C:N and moisture balance can be maintained, virtually any organic material can be composted. If any of the following can be obtained without transportation expenses, add them to your compost pile — sawdust, nursery wastes, fruit and vegetable residue from processing plants, feathers, grass and lawn clippings, vegetable market wastes, garden wastes, leaves, wood shavings and even seaweed. But, before using such materials, satisfy yourself that they are uncontaminated by heavy metals or others toxins.

Adding clay soil is also a good idea. The clay will help reduce nitrogen losses by holding any liberated ammonia within the heap until the micro-organisms can stabilize the volatile nitrogen.

EQUIPMENT NEEDED

The unique conditions on each farm means that giving hard and fast recipes for successful composting is not a good idea. Every farmer should be prepared to experiment with materials and techniques. The following methods work successfully for many farmers and are offered here as guidelines. Use a front end loader to remove the solid manure from where it is stored and to load the manure spreader. The manure spreader should be a "power-take-off" type so that it can discharge its load while parked.

BUILDING A WINDROW

The site chosen for building the compost piles or windrows should not be near a waterway or tile drains. Ideally, there should be a way to retain the liquids that will leach out of the pile and seep into the soil if the pile becomes too wet when rainfall is high. This can be done if the site selected is on a slight slope, with a receptor pit to catch the run-off. Otherwise, use level land and do not use the same place each year. Heavy clay soil is ideal because it prevents any leachate from reaching the water table.

The tractor pulls the spreader, parks and unloads the manure through the beaters which shred the manure and add air as the pile is being built. The spreader is then pulled ahead about 1 metre (3 feet) and the procedure is repeated. This method is used to form the windrow and for the turning of the pile.

Windrows created from cattle manure should be made about 1.25 metres (4 feet) high and 2-2.5 metres (6-8 feet) wide. Horse manure can overheat so windrows should only be 0.5 metre (1.5 feet) high for efficient cooling. However, because of the large surface area, shallow windrows can dry out quickly in drought conditions or suffer from leaching in heavy rainfall; both conditions require management correction. It is more practical to layer horse manure as it comes out of the barn and add further layers every few days. Hog manure is very dense and large quantities of straw need to be added; otherwise it will tend to decompose anaerobically.

Any liquid manure that seeps out of the windrow should be pumped back onto the drier materials and allowed to percolate through.

Manure from loose-housing bedding packs may overheat due to the large straw:manure ratio. Limiting the amount of air in the pile will prevent this. This can be done by using a dump trailer to create windrows instead of a manure spreader. Air can also be expelled by tramping down on the pile.

TEMPERATURE

Microbial activity generates heat and the pile can warm to about 60-70°C within one week if the hot composting method is used; then decrease over a few weeks. To keep the temperature at 40-50°C and reduce the loss of nutrients, manure should be stored so that some decomposition takes place before it is put into windrows. Adding soil will also keep the temperature lower. Overheating stops microbial action and causes excessive nutrient loss, hence temperatures over 60°C should be avoided. Temperatures above 55°C may be desirable if there are disease problems in the barn. The temperature in the pile can be monitored with a temperature probe 0.5-1 metre (2-3) feet long and corrective action taken if necessary. Overheating may indicate deficient moisture levels, too much N or too much C. In cold weather, warmer conditions can be maintained by covering the pile with black plastic. This will also prevent nutrient loss by leaching.

TURNING THE PILE

Turning the pile is not needed if optimum conditions are met. Some people advocate turning the pile to speed the decomposition process and obtain mature compost in about 10 weeks. This, however, will also likely cause high nutrient losses; turning may be required to improve conditions. The simplest method is to use a front-end loader to push the piles over and reform them. Using the manure spreader will do a better job of remixing the compost. Commercial-brand compost turners are available but are not essential for making good compost.

MOISTURE AND AERATION

The moisture content of the pile will often determine if turning is necessary; it should be about 50 per cent. If a moisture meter is not available this can be tested by squeezing the material in your hand — if it glistens and small moisture droplets appear, the moisture content is sufficient. Beginners at composting tend to have piles that are either too dry or too wet.

If the pile is too moist, water replaces the air in the pile leading to anaerobic conditions. Turning the pile to reintroduce air changes the pile from an anaerobic to an aerobic system. The smell of the compost should be your guide; it should be sweet-smelling. An unpleasant smell indicates that anaerobic decomposition is taking place. If you are in a wet area, either build a roof over the pile or cover the pile with straw or black plastic to avoid leaching of potash and trace elements. On the other hand, if the pile is too dry, biological activity will cease. In this case, water will have to be added. This is best done when the compost is being turned.

MINERAL ADDITIONS

Mineral additions to the compost, in the form of powdered rock, can be useful. Rock phosphate is often added to the gutter in the barn to reduce immediate nitrogen loss and manure odor. It has the added benefit of increasing phosphorus reserves and encouraging nitrogen fixing, blue-green algae. Clay minerals such as bentonite and montmorillonite are added at any time. They trap liberated ammonia, increase the cation-exchange capacity of the compost and thus make the nutrients more available to the crops and hasten the formation of the clay-humus complex. Small amounts of kelp products or other trace element sources can be added to the compost directly.

APPLICATION OF COMPOST

Compost piles or windrows take about three months to mature and should be black-brown in appearance with a crumbly texture. Ideally, it should be spread as soon as possible after it is finished. The longer it sits, the more it mineralizes and loses available nutrients. The presence of weeds on the pile indicate that mineralization is occurring and that the compost should have already been applied to the land. Application rates vary depending on the crops, the needs of the soil and the age of the compost but an application of at least 15 t/ha is usually recommended at some point in the rotation.

Applying compost to crop land before spring planting may encourage the growth of weeds. If composting is started just after seeding it should be ready to apply to stubble fields after harvest. Incorporate the compost lightly into the top 10 cm of the soil; if any deeper, much of the nutrient gains and soil conditioning value will be wasted. Cover crops can be seeded immediately after applying the compost. Compost should, however, be applied to growing plants whenever possible; it will not burn them and the nutrients are readily utilized.

As mentioned, young compost is high in soluble nutrients, whereas mature humus has a higher proportion of stable humus. This will determine where you should apply your compost. Crops that are nutrient-demanding, such as winter wheat and corn, should receive applications of young compost. Mature compost applied to these crops will have little immediate benefit. During the transition period some compost should be applied to all the fields to stimulate microbial activity.

LIQUID MANURE SYSTEMS

Many pig, poultry and dairy operations have slurry or liquid manure systems. Raw slurry can lead to serious pollution problems, by fouling waterways and disrupting the flora and fauna in the soil.

A form of composting can be done with liquid slurry by microaerating the slurry in the tanks. The aim of microaeration is to reduce nutrient loss from volatilization which occurs when nitrogen changes into gaseous ammonia. The advantage of this microaeration process is that it cuts down on odors from anaerobic conditions which result in the formation of methane and hydrogen sulphide. This significantly reduces weed seed and pathogens and also reduces nitrogen loss because the nitrogen is converted from ammonia to bacterial protein — a stable form of nitrogen. Application of liquid manure is most beneficial if applied onto green manure crops in late summer or fall. This is when the ground is dry and the green manure crop can efficiently tie up the nutrients.

UTILIZATION OF ORGANIC WASTES: ON-FARM COMPOSTING

BENEFITS OF USING COMPOST

Compost can provide valuable nutrients and organic matter to soil, depending upon the feedstocks (raw materials used) and upon compost management. A chemical analysis of a representative sample of a compost will indicate its total nitrogen, available nitrogen, phosphorus, and potassium. Most composts contain relatively low concentrations of one or more nutrients and are not necessarily considered good "fertilizers"; however, as soil amendments, they are good sources of organic matter.

Nitrogen and phosphorous in compost are generally found in both plant-available forms (NO_3, NH_4, and P_2O_5) and organic forms. Much of the nutrients bound in organic forms will be made "plant-available" as the organic matter decomposes. Therefore, readily available nutrients in compost can be much lower than in raw waste, but a "timed-release" effect occurs in the later, slow-release of nutrients "bound" initially in organic forms.

Why should you consider making compost?

The principal advantages in making your own compost are (1) composting is a good treatment option for many agricultural wastes and can produce a marketable product; and (2) composting stabilizes organic materials so they can be stored safely, transported easily, and applied at a convenient time.

Large-scale composting is practiced by municipal wastewater treatment plants, agricultural producers, industrial waste generators, and commercial composters (who are in the business of composting wastes collected from various sources). The compost product is often marketed to "wholesalers" (nurseries, farms, landscaping companies, etc.) or is packaged for retail sale to the general public. Depending upon compost production and markets, prices of commercially available compost may vary considerably.

Raw waste materials may be applied directly to farmland, but there are limitations to the timing of these applications. In fact, when a producer is ready to apply manure to his fields, so are most of his neighbors. Animal waste generators may find themselves "buried" with manure during the winter, and by spring planting be "sold out" of the commodity. If a farmer could obtain the waste when it is abundant, he may be able to pay a lower price. However, storing large quantities of manure can bring negative effects, including odor problems.

Compost is easier to handle and store than raw waste. It does not have an offensive odor and is less likely to contribute to water contamination. A farmer who composts waste materials will likely have greater flexibility; his soil amendment will be available at the time he is ready to use it. The following information may help you decide whether composting is appropriate for your operation.

What you will need to begin composting?

The feasibility of composting in your operation depends upon availability of materials, equipment, and labor. Of course, the compost process cannot occur without "feedstocks," or raw waste materials to be composted.

Basically, composting is a biological treatment of organic wastes in which microorganisms (bacteria, fungi, and actinomycetes) consume and "break down" organic material into more stable forms. Simply put, we are feeding these organisms, and to make them work efficiently, we attempt to provide them with a balanced diet and a comfortable environment.

To provide nutrients for composting microorganisms, you will need a nutrient-rich feedstock. Animal manures are good sources of macronutrients and micronutrients. Manure can be composted by itself, but it may be too "compacted" for good aeration, and excess nitrogen can be lost through volatilization of ammonia (thereby producing a strong odor). A carbon-rich, relatively low-nutrient feedstock can be mixed with the nutrient-rich feedstock, providing a better "food" for the microorganisms. A carbon:nitrogen ratio between 20:1 and 30:1 is a good target. Since some feedstocks decompose more slowly, this ratio may be finetuned to account for relative availability of carbon and nitrogen.

Dense feedstock mixtures may not have sufficient aeration and become anaerobic (subsequently producing offensive odors and potentially harmful gases). Bulky material, such as wood chips or straw, can be added to reduce the bulk density and increase air volume in the compost pile. Bulking agents can be left in the compost, or they can be screened out and re-used in future compost piles.

Availability of feedstocks is critical to your decision about whether to compost. If you produce one or more of the ingredients within your farm operation, you may be able to obtain the remaining feedstocks and still operate economically. It is important to take transportation costs into consideration. It is not economical to haul feedstock or compost long distances. Consider other producers and industries in your area. Perhaps you can organize a cooperative composting effort, or perhaps you can charge a tipping fee to accept waste from other sources.

You also will need space to build the compost piles, store feedstocks, and maneuver equipment. If possible, locate the composting operation downwind of residences and away from surface water and wells. Be sure to consider your neighbors' feelings toward your composting activity before you begin operations. Odors and flies from stored feedstocks and compost piles, and dust and odors associated with windrow turning operations can be offensive to some neighbors.

The main equipment needed for composting on a moderate to large scale is machinery—to construct, mix, and move material in the compost pile or windrow. A front-end loader and a truck may be all you need. Other equipment to consider may include (1) chipping or shredding equipment (depending upon the feedstock); (2) a windrow turner (available in many sizes and prices); (3) screening equipment (if you wish to recycle bulking agents); (4) aeration equipment ; and (5) composting thermometer or temperature probe (to monitor progress of your compost). You will also need to take labor requirements into consideration. Do you have time to build the windrow and mix it as needed?

How to compost?

An excellent practical reference for composting operations is the *On-Farm Composting Handbook* (Northeast Regional Agricultural Engineering Service[NRAES], 1992). It addresses composting benefits and drawbacks, methods and materials, management, equipment, and marketing of the final product. This inexpensive reference is available from NRAES or from the state extension agricultural engineering specialist.

Composting is a natural process, not an exact science. Therefore, a lot of flexibility exists in developing a composting system. Management of a composting operation may be directed toward one or more of the following objectives.

1. *Minimizing time required for composting:* If space is limited, an operator will want to maximize the processing rate or minimize the composting area required. Generally, aerobic composting is much faster than anaerobic processes; therefore, more material can be processed on a given area in an aerobic composting system.
2. *Maximizing destruction of pathogens and pests:* High-temperature aerobic composting destroys weed seeds, insects, and pathogens; therefore, temperature may be a primary concern. Regular mixing of the material will facilitate aerobic conditions and ensure that all feedstock material is exposed to high temperatures.
3. *Minimizing nuisance conditions:* Aerobic composting is relatively odor free, compared to anaerobic processes. Even in aerobic systems, however, excessive nitrogen in the feedstock mixture can be volatilized, thereby generating odors.
4. *Minimizing labor costs:* Static piles (not turned after construction) and anaerobic systems require less handling than other systems. Various levels of automation can reduce labor requirements. Of course, automated systems may require high initial capital outlays for equipment and/or high energy costs.
5. *Optimizing the final product quality and marketability:* Managing nutrient mixtures and feedstocks can result in a compost product tailored for a specific market.

Generally, composting feedstocks are mixed mechanically with front-end loaders or windrow turners. The materials are placed into piles or windrows that should be sufficiently large to generate and store heat to promote high-temperature composting, but not so large that the materials become excessively compacted. Typical windrows range in size from 4 to 8 feet tall and 10 to 20 feet wide at the base. The dimensions should be no larger than available equipment can accommodate.

Windrows may be aerated through passive aeration (natural convection) or forced aeration (by fans or compressed air). They may be static piles or turned regularly (every three days to two weeks). Management method depends upon characteristics of the waste, labor and equipment available, and operator preferences.

A carbon:nitrogen (C:N) ratio of between 20:1 and 30:1 is generally considered appropriate for agricultural wastes. Higher C:N ratios slow material decomposition, because low nitrogen limits microbial activity. Lower C:N ratios may contain excessive nitrogen that may be volatilized as ammonia, thereby producing odors and wasting nitrogen. If odors are a major concern, consider a feedstock mixture with a higher C:N ratio.

Moisture content between 40% and 60% is a good target range. Moisture is needed for microbial activity, but excessive moisture inhibits gas exchange and may result in anaerobic conditions. The compost mixture should feel moist to the touch, but not be soupy. Very wet feedstocks may be dried before mixing, or a dry bulking material can be used to absorb moisture. Consider protecting the compost piles from excessive rainfall or ponded water. Some moisture will be removed from the mixture during the composting process. During dry weather, the mixture may need water added to maintain moisture.

Nutrients are required for microbial activity. Most agricultural wastes and municipal sewage sludges contain sufficient macronutrients and micronutrients to support decomposition. Some industrial wastes may be deficient in one or more nutrients.

Feedstock mixture pH can be important. Microorganisms tend to modify their environment, and products of decomposition may alter pH over time. Near-neutral pH is preferred for most efficient microbial activity. Some food processing wastes and industrial wastes may exhibit levels of alkalinity or acidity that inhibit nutrient availability or microbial activity. Chemical analyses of material samples will indicate whether pH or nutrients need to be adjusted.

Bulk density should be low enough (less than 40 lb/ft^3) to allow for good aeration. Dense manures and sludges can be "lightened" by adding of bulking agents, such as wood chips, corn cobs, and straw.

Insulation material can be used if cold weather keeps compost temperatures down. It also can help reduce odor emissions from a pile. Preferred insulation materials include finished (recycled) compost and/or bulking materials.

Odor should not be a problem in a well-managed composting operation. If odors are present, the mixture should be turned to improve aeration.

Adding a carbon-rich bulking agent may be helpful to improve the C:N ratio and aeration. Excessive moisture may contribute to odor problems; therefore, management of the process may require shelter and drainage during wet weather. Odors can be generated from stored feedstocks prior to mixing; if odorous materials (e.g., seafood wastes) are to be used as feedstock, they should be mixed immediately upon receipt. Consider having nonodorous bulking agents available before odorous materials are delivered in order to avoid unnecessary delays in mixing.

SOME POSSIBLE FEEDSTOCK COMBINATIONS

Almost any organic material can be composted. The main objectives of C:N ratio, moisture content, and bulk density can be achieved with a variety of feedstock combinations. Therefore, gardeners and farmers alike often can easily identify likely "recipes" from materials on-site. Some suggestions include:

I. Combinations of poultry litter with bedding material and additional carbon-rich bulking materials, including (1) broiler litter containing wood shavings as bedding material composted with peanut hulls; and (2) broiler litter containing wood shavings as bedding material composted with shredded pine bark.

II. Municipal biosolids composted with combinations of sawdust, yard wastes, bark, vegetable trimmings, animal bedding and manures.

III. Ground (shredded) yard wastes, dairy manure, and food processing wastes.

To determine an appropriate ratio of feedstocks for your system, determine carbon and nitrogen contents, moisture content, and bulk density for each feedstock. Design your mixture by balancing the C:N ratio with your principal feedstocks; then adjust moisture and bulk density with bulking agents and/or water.

Note that feedstocks can decompose at different rates. For instance, broiler litter may decompose much faster than sawdust in a particular mixture. Therefore, the sawdust may not contribute carbon at a rate sufficient to balance excess nitrogen in the broiler litter. The C:N ratio should be used to generate an initial "recipe," which will be adjusted later to accommodate specific conditions.

INOCULANTS

Some commercially available inoculants are marketed to improve compost processing. While these materials are unlikely to be harmful, generally they are not necessary. They may improve decomposition of some relatively sterile industrial wastes, and they may even speed the heating cycle of the first compost "batch" for an operation. However, most agricultural wastes contain sufficient microbial populations to initiate composting.

Perhaps the best way to inoculate a pile is to mix some compost from the previous batch into feedstock of the new batch. This is especially effective if the feedstocks are similar in the different batches.

What you should expect?

Once a windrow is constructed, the microbial activity will generate carbon dioxide, water vapor, and heat. This activity will be evident through increased pile temperature and possibly by visible "steam" coming from the pile. Over time, some of the carbon and water will be removed, thus causing the pile to lose mass (size) and moisture. If water content falls below 40% to 50%, water should be added and mixed into the composting feedstocks. Gradually the C:N ratio will fall as the readily compostable carbon is metabolized by microorganisms, and the nitrogen is converted to nitrate and organic forms.

Pile temperature is a good indicator of the compost process. During the first few days of composting, pile temperature should increase to between 105 °F and 149 °F. The high temperature may be maintained for several days, until the microorganisms begin to deplete their food source or until moisture conditions become less than optimal. Mixing the composting feedstock brings more undecomposed "food" in contact with the microorganisms, replenishing their energy supply. If moisture content in the mixture is too high, insufficient aeration will limit activity of high-temperature aerobic microorganisms. If the moisture content is too low, it will limit microbial activity. Once the optimum moisture level is restored and the feedstocks have been remixed, temperature will increase again. The heating-cooling-remixing cycle is repeated until the readily decomposable material is depleted.

How to know when it is "done"?

After the readily decomposable material is depleted, the compost pile will no longer heat upon remixing. The temperature will continue to drop to ambient. Only very slow decomposition will continue. The material should have a pleasant "woodsy" odor and a friable texture similar to a good potting soil. The material will likely feel moist and cool and have a dark brown color. Several tests can be used to determine "doneness" of the compost, including incubation to test for generation of metabolism by-products and respirometer testing to measure oxygen use. Often it is recommended that compost "cure" for several months to allow for continued slow decomposition of more resistant constituents.

TROUBLESHOOTING

Some problems encountered in a composting operation may include odors, temperature abnormalities, and fly or mosquito problems. Odor problems, addressed throughout this article, may result from odorous feedstock, excess nitrogen released in ammonia form, and odorous by-products (such as hydrogen sulfide gas) of anaerobic decomposition. Control these with proper storage or timely use of feedstock, sufficient carbon sources and bulking agents to control ammonia release, and appropriate moisture and aeration to ensure aerobic conditions. An acceptable temperature can be maintained by building piles or windrows of appropriate size, maintaining moisture content between 40% and 60%, and mixing properly. Insect problems are generally associated with exposed wet feedstock or standing water on site. Timely mixing of feedstocks and good control of water from drainage, runoff, and leaching will limit opportunity for insect breeding.

UTILIZING COMPOST

High-quality compost can provide soil nutrients for improved crop, garden, and/or turf productivity. Even compost relatively low in available nutrients can provide valuable organic matter to soil, thereby improving permeability and water-holding characteristics. Cured compost can be marketed as an amendment to improve soil productivity, as a component in potting soil, or as a mulch for landscaping, gardening, etc. If high temperatures have been maintained in the composting process, the compost will be relatively free of weed seeds, insects, and pathogens compared to untreated organic wastes.

NUTRIENT CONTENT OF THE ORGANIC WASTES

The amount of nitrogen found in animal manure depends upon the age and type of animal, feeding rate, type of ration, and how the waste was stored and handled both before and after being applied to the soil. To determine the amount of nitrogen applied, a farm operator must know both the amount of waste applied and its nitrogen content. A laboratory can analyze the manure for total nitrogen content, and the amount of waste applied can be determined by using a calibrated application system.

RATE OF MINERALIZATION

Before nitrogen and other nutrients in manure can be used by plants, soil microbes must break them down into forms that are readily available to plants. This process is called "mineralization."

The rate of mineralization is called a "decay series." Different types of waste will have different rates of decay. A decay series estimates the percentage of mineralization that will occur in the years following a manure application. For example, a decay series of 0.35, 0.15, 0.10, 0.075 means that following a dry corral manure application, 35% of the nitrogen is mineralized the first year, 15% of the residual (that which was not previously mineralized) is released in the second year, 10% in the third year, and so on. If the nitrogen concentration of a waste material and its decay series are known, the amount of waste needed each year to supply a constant amount of nitrogen can be calculated. The approximate application rates of five waste materials needed to maintain an annual mineralization rate of 200 pounds nitrogen per acre.

Table. Input of five manure types needed to maintain annual mineralization rate of 200 pounds nitrogen per acre.

Material and Decay Series	Pounds of N/ton	Annual application rate Year				
		1	2	3	4	5
		tons/acre				
Poultry manure, 1.6%N 0.90, 0.10, 0.075, 0.05	32	6.9	6.2	5.7	5.4	5.5
Fresh bovine waste, 3.5%N 0.75, 0.15, 0.10, 0.075	70	3.8	3.0	2.7	2.5	2.4
Dry corral manure, 2.5%N 0.40, 0.25, 0.06, 0.03	50	10.0	3.8	6.2	4.8	5.8
Dry corral manure, 1.5%N 0.35, 0.15, 0.10, 0.075	30	19.0	10.9	9.0	8.0	10.8
Dry corral manure, 1.0%N 0.20, 0.10, 0.075, 0.05	20	50.0	25.0	18.8	18.5	27.5

WEED SEEDS

Manures can contain seeds from weeds that can prove difficult to control. Because heat generated in manure stockpiles decreases the viability of weed seeds that may be present, the use of well-aged manures instead of freshly excreted materials will help reduce the likelihood of weed infestations from manure applications. Careful attention to the origin and quality of animal feedstuffs may also help reduce the severity of manure-transmitted weed problems.

METHOD OF APPLICATION AND TIMING OF INCORPORATION

Animal manures and wastes should be injected or uniformly broadcast on cropland at recommended rates and then incorporated into the soil as soon as possible. Plowing or rototilling of the soil following surface applications of manure is recommended. Subsurface injection of fluid materials generally does not need additional tillage operations.

Immediate mixing with the soil will greatly reduce odor, nitrogen losses due to ammonia volatilization, and the potential for groundwater and surface water contamination resulting from runoff. The effect of an increasing time lag between surface application and incorporation of animal manure.

Table. The effect of time lag between surface application and incorporation of poultry and other manures on the percentage of manure nitrogen available to crop plants.

Time of incorporation	Percentage of manure nitrogen available	
	Poultry	Other
Immediate	75	50
After 2 days	45	35
After 4 days	30	30
After 7+ days	15	20

Note:

The information found in this fact sheet was excerpted from *Nitrogen Fertilizer Management in Arizona,* May 1991. University of Arizona Cooperative Extension.

SALT CONTENT

Manures from concentrated animal feeding operations are usually high in salt content. Most dairy and feedlot manures contain 5 to 10% salt (50,000 to 100,000 ppm). Frequent and/or large (20 tons per acre) applications of manure to cropland increases the risk of salt injury to plants. Salt-sensitive plants such as lettuce, tree fruits and nuts are especially susceptible. The following management practices are recommended for applying animal manures to cropland:

1. Use well-aged manures rather than fresh manures taken directly from feedlots;
2. Apply up to 5 tons per acre of dry matter per year or 10 tons per acre every other year;
3. Use supplemental nitrogen fertilizers only as required based on tissue tests, plant performance and previous experience;
4. Plow or rototill manure into the soil, irrigate and wait at least 30 days before planting;
5. Do not apply manure where water penetration is poor; and
6. Monitor soil salinity and sodium levels by periodic soil tests.

AGRICULTURAL APPLICATION OF BIOSOLIDS

Biosolids are another type of organic waste that can be applied to cropland. The following guidelines are recommended for applying biosolids:

1. Make sure that the soil pH is above 6.5 at the time of application to minimize leachability of trace metals and their uptake by plants;
2. Observe the cumulative loading limits for soil applications. No more than the specified total amount of each pollutant may be applied to a given site in perpetuity;
3. Use crops which exclude heavy metals from the entire plant or from harvested plant parts (e.g., grain, seed or fruit crops);

4. Use sound soil management practices to reduce runoff and erosion; and
5. Monitor toxic element applications and accumulation in soil and plant tissue using periodic laboratory tests.

Table. The maximum amount of pollutants that can *ever* be applied to a given site.

	Cumulative Pollutant Loading Rates (dry weight)	
	pounds/acre	kilograms/hectare
Arsenic	36.6	41.0
Cadmium	34.8	39.0
Copper	1,339.3	1,500.0
Lead	267.9	300.0
Mercury	15.2	17.0
Nickel	375.0	420.0
Selenium	89.3	100.0
Zinc	2,500.0	2,800.0

REGULATIONS FOR THE APPLICATION OF BIOSOLIDS IN ARIZONA

The application of biosolids to cropland is also regulated by law but with a different set of requirements than for animal manure application. The following list highlights regulations relating to the application of biosolids that are *not*exceptional quality to cropland. For further information as well as permit requirements, please contact the Arizona Department of Environmental Quality (ADEQ).

Permit: Applicators must file a "Request for Registration" form with ADEQ prior to receiving biosolids.

Proximity to wells: Biosolids cannot be stored or applied closer than 1,000 feet from a public or semipublic drinking water well and no closer than 250 feet from any other water well.

Application site: Biosolids should be applied to soil with a pH of 6.5 or greater. The site should have a slope of less than 6% and be at least 32.8 feet from surface water.

Depth to groundwater: This depends on the pathogen reduction class of the biosolids. For Class B biosolids, the minimum depth to groundwater is 10 feet; 40 feet if soil is gravel or coarse to medium sand. For Class A biosolids, the minimum depth to groundwater is 5 feet.

How biosolids are applied: Biosolids cannot be applied at an application rate greater than the agronomic rate of the crop.

GREEN MANURESGREEN MANURES

Despite special efforts at increasing the supplies of farmyard manure and compost, the supply of farm and other organic manures is scarce and ever becoming more costly. Green-manuring, wherever feasible, is the principal supplementary means of adding organic matter to the soil. It consists in the growing of a quick-growing crop and ploughing it under to incorporate it into the soil. The green-manure crop supplies organic matter as well as additional nitrogen particularly if it is a legume crop, which has the ability to acquire nitrogen from the air with the help of its root-nodule bacteria. A leguminous crop producing 8 to 25 tonnes of green matter per hectare will add about 60 to 90 kg of nitrogen when ploughed under. This amount would equal an application of three to ten tonnes of farmyard manure on the basis of organic matter and its nitrogen contribution. The green manure crops also exercise a protective action against erosion and leaching.

The crops most commonly used for green-manuring in this country are the following:

Sunnhemp (Orotalaria juncea), dhaincha (Sesbania aculeata), cluster-bean (Cyamopsis tetragonoloba), senji (Melilotus parviflora), cowpea (Vigna catjang, V. sinensis), horse-gram (Dolichos biflorus), pillipesara (Phaseolus trilobus), berseem or Egyptian clover (Trifolium alexand-rinum). Lentil (Lens esculenta) is recommended in Kashmir for green-manuring paddy. Sown in late autumn, it is said to provide a winter cover and make new growth in early spring for ploughing under before the sowing of paddy. Sunnhemp is the most outstanding green-manure crop. It is well suited to almost all parts of the country and fits in well with sugarcane, potatoes, garden crops and the second-season paddy in southern India and with irrigated wheat in the north. Dhainchais in wide use in Assam, Bengal and Tamil Nadu. It does well on alkaline and waterlogged soils. Cluster-bean, berseem and senji do well in Punjab, Uttar Pradesh, Rajas than, Delhi and some parts of Madhya Pradesh. Berseem is well suited for orchards and the irrigated crops of cotton and sugarcane sown in spring or early summer, as in Punjab and Uttar Pradesh. Cowpea and horse-gram are used for green locality is naturally the one most suited to its soil and climatic conditions.

Very often, berseem, senji and lucerne (Medicago sativa), and sometimes sunnhemp are grown partly for fodder and partly for fodder and partly for green manuring. In the case of annual crops of senji and berseem one or more cuttings are taken for use as green-foedder. Lucerne which is allowed to grow for two or three years, is cut seven to eight times for the same purpose. In the case of sunnhemp, the tops are fed to the cattle. In all these instances, the residues (roots and stumps) are incorporated into the soil. These crop residues contain considerable amounts of nitrogen, phosphorus, potassium and other mineral nutrients, besides organic matter. In the case of orchards, the annual green manure-cum-forage crop should be grown at such a time as to interface the least with tree growth and fruit development.

Pulses form an essential part of the India diet, and are grown commonly as pure crops in rotation or mixed with cereals, oilseeds, and fibre crops. Roots and stubble of these pulse legumes return to the soil small quantities of organic matter rich in nitrogen. The inclusion of groundnut on the cotton-jawar rotation of centra, southern and western India, the growing of a quick-maturing variety of ming (Phaseolus aureus) before wheat in Uttar Pradesh, and wheat and rabi jowar in Marathwada Division of Bombay the sowing of peas in the standing crop of irrigated cotton in Uttar Pradesh, and the sowing of sunnhemp in the standing paddy crop in Andhra Pradesh, Tamil Nadu and Karnataka states, or of val IDolichos lablab) in the coastal paddy areas of Maharashtra, are valuable practices for soil improvement. All these leguminous crops leave the soil in a better physical conditions and richer in nitrogen.

The growing of pulses mixed with cereals all over India, the intercropping of cotton with groundnut of with tur (Cajanus indicus) in central and southern parts of the country, or with cluster-bean, mung and moth (Phaseolus aconitifolus) in Punjab and adjoing states; the growing of wheat mixed with peas and gram in northern and central India; and the growing of fodder sorghum mixed with Dolichos lablab in some parts of Tamil Nadu also enrich the soil.

In localities near forests in Tamil Nadu, Karnataka and Andhra Pradesh, the paddy crop is often manured with green forests leaves. These are incorporated into the soil at the time of puddling. In recent years, extensive efforts have been made in these states to plant Glyricidia maculata and Sesbania speciosa on the borders of paddy fields or in other vacant spaces to provide green leaf for manuring the paddy crop. Grown from seedlings or rooted stumps, planted 2 m apart, each Glyricidia plant is said to give annually tow cuttings each of about 6 to 12 kg of green leaf. It does well in both red black soils. Similarly Sesbania speciosa seedlings planted 10 cm apart on paddy borders produce 1,000 to 2,500 kg of green leaf for manuring 0.4 ha of paddy. Only 115 g of seed is needed to provide seedlings sufficient for border-planting of two hectares of paddy. In certain other paddy-growing areas, Pongamia pinnata (karanji), Tephrosia, Terminalia and other trees yielding large quantities of leaves are planted for use as a green manure. In the Malabar districts of Tamil Nadu, Indigofera teysmanni is grown to provide green leafy twigs for manuring paddy.

NITROGENEOUS FERTILIZERS

According to the manner in which their nitrogen is combined with other elements, the nitrogenous fertilizers are divided into four groups; nitrare, ammonia and ammonium salts, chemical compounds containing nitrogen in the amide form, and plant and animal by-products.

SODIUM NITRATS

It is also known as 'Chilean' nitrate. It owes its importance as a pioneer nitrogenous fertilizer. It occurs in natural deposits in northern Chile and is refined before shipping. The refined product contains about 16 per cent nitrogen in the nitrate form, which renders it directly available to plants. For this reason, it is highly valued as a source of nitrogen when applied as top-and side-dressings, especially to young plants and garden vegetables, which need readily available nitrogen for quick growth leached out from the soil. For wheat, maize, barley, cotton, sugarcane, etc., it is as beneficial as ammonium sulphate.

Sodium nitrate is particularly useful for acidic soils. Its continued and abundant use in soils is said to cause deflocculation and develop a bad physical condition in regions of low rainfall. It should be stored in a dry warehouse

AMMONIUM SULPHATE

It is the most widely used fertilizer in the country. It is a white crystalline salt, containing 20 to 21 per cent nitrogen. It is easy to handle and it stores well under dry conditions. During the rainy season, it sometimes, forms lumps. These lumps should be powdered before use. Being soluble in water, it acts quickly, but despite its high solubility, its nitrogen is not readily lost in drainage, because the ammonium ion is retained by the soil particles. It is, therefore, very suitable for wet-land crops, *e.g.,,* paddy and jute. It has also been found useful on wheat, cotton, sugarcane, potatoes and many other crops grown on a wide variety of soils. It has. however, an acid effects on the soils. Its long-continued use increases soil acidity and lowers the yield. The application of this fertilizer to acid soils improved the yield of tea plants considerably. It is advisable to use this fertilizer in conjunction with bulky organic manures to safeguard against the ill effects of continued application of ammonium sulphate to field and horticultural crops. Ammonium sulphate can be applied before sowing, at sowing time, or as a top-dressing to the growing crop. It should not be applied along with, or too close to, the seed, because in concentrated form, it affects seed germination very adversely.

AMMONIUM NITRATE

It is a white crystalline salts, containing 33 to 35 per cent nitrogen, half as nitrate nitrogen and half in the ammonium form. In the ammonium form, it cannot be easily leached from the soil. This fertilizer is quick-acting, but highly hygroscopic and not fit for storage. Granulation of the material and a light coating of the granules with oil reduce hygroscopicity to some extent. It has an acidulating effect on the soil. Under certain conditions, it is explosive, and, therefore, it should be handled cautiously.

'Nitro Chalk' is the trade name of a product formed by mixing ammonium nitrate with about 40 per cent limestone or dolomite. It is granulated, non-hazardous and less hygroscopic. It contains 20.5 per cent nitrogen, half in the form of ammonia and half as nitrate. The presence of lime in it makes it particularly useful for acid soils.

AMMONIUM SULPHATE NITRATE

It is a mixture of ammonium nitrate and ammonium sulphate. It is available in a white crystalline form or as dirty-white granules. This fertilizer contains 26 per cent nitrogen, three-fourths of it in the ammoniacal form and the rest (6.5 per cent) as nitrate nitrogen. It is non-explosive and not as deliquescent as ammonium nitrate. It is readily soluble in water and is very quick-acting. Its keeping quality is good and it is useful for all crops. Its keeping quality is good and it is useful for all crops. Its acid effect on the soils is only one-half of that of ammonium sulphate. It can be applied before sowing, at sowing time or as a top-dressing, but it should not be applied along the seed.

AMMONIUM CHLORIDE

It is a white crystalline compound, possessing a good physical condition and containing 26 per cent ammoniacal nitrogen. It is extensively used on paddy in Japan, In India, it is used largely in industries. In general, it is similar to ammonium sulphate in action. It is usually not recommended for tomatoes, tobacco and such other crops as may be injured by chlorine.

Urea

It is a white, crystalline, organic chemical. It is a highly concentrated nitrogenous fertilizer, containing 45 to 46 per cent of non-proteined organic nitrogen. It is fairly hygroscopic and presents considerable difficulty, it is also produced in granular or pellet forms and is coated with a non-hygroscopic inert material. It is highly soluble in water and, therefore, subject to rapid leaching. It is, however, quick-acting. When applied to the soil, its nitrogen is rapidly changed into ammonia. Like ammonium nitrate, urea supplies nothing but nitrogen. It may be applied at sowing time or as a top-dressing, but should not be allowed to come into contact with the seed. It is suitable for most crops and can be applied to all soils.

Ammonia

It is a gas containing about 80 per cent of nitrogen. Under suitable conditions of temperatures and pressure, it becomes liquid (anhydrous ammonia). Another form, 'aqueous ammonia', results from the absorption of ammonia gas into water, in which it is soluble. Ammonia is used as a fertilizer in both these forms. Anhydrous ammonia can be applied by introducing it into irrigation water, or directly into the soil from special containers, which makes its use rather expensive. Its possibilities as manure for paddy, sugarcane and cotton are being investigated at Bangalore in the Karnataka state. In manurial experiments on cotton in Maharashtra, aqueous ammonia has been found to be as efficient as ammonium sulphate.

Calcium Ammonium Nitrate

Calcium ammonium nitrate is a fine free-flowing, light brown or grey granular fertilzer. It is commercially prepared from ammonium nitrate and ground limestone. It is almost neutral and can be safely applied even to acid soils. Its total nitrogen content may vary from 25 to 28 per cent. Half of this total nitrogenis in the ammoniacal form and half is in nitrate form. According to the prescribed standards, its moisture content should not be more than 0.5% by weight. As regards the particle size, 90 per cent of the material should pass through 4 mm IS sieve and be retained on a 1-mm IS sieve but not more than g per cent shall be below 1 mm.

GREEN MANURING IN RELATION TO THE SOIL FERTILITY AND SOIL HEALTH

Owing to the constant production of crops from the soil, the latter is being depleted gradually of its nitrogenous and other nutrients. An ordinary crop takes about 25 lb of nitrogen from an acre. It is, therefore, necessary to replenish the soil with the elements, which are removed by the crops year after year. Organic matter is the life of the soil because it contains all the essential elements required for plant growth. It also serves as food for soil bacteria. Decomposed organic matter, known as humus, improves the soil tilth and helps the plant to grow. Well-stored farmyard manure is most important of all organic manures, but it is not available in sufficient quantity. Farm yard manure, if not properly stored, loses its nutrient-supplying value to a great extent. Therefore, in order to conserve farmyard manure and town refuse properly, two schemes were taken in hand by the Agricultural Department during the Second Five Year Plan. But the answer to this problem is green manuring in which no such losses are there.

GREEN MANURING

Farmyard manure and compost are not available in sufficient quantities to the farmers to meet their full requirements. Artificial fertilizers are also in short supply. Owing to the intensely hot summers, the available humus in the soil is burnt up quickly. A periodical application of organic matter is, therefore, essential to replenish the loss of humus, which is necessary for keeping the soil in good condition by enhancing the supply of nitrogen and by promoting the growth of microorganisms. A leguminous crop producing 8 to 25 tonnes of green matter per hectare will add about 60 to 90 kg of nitrogen when ploughed under. This amount would equal an application of three to ten tonnes of farmyard manure on the basis of organic matter and it's nitrogen contribution. The green manure crops also exercise a protective action against erosion and leaching of the various nutrients into the deep soil layers. Green-manuring is, thus, a very useful soil-improving practice for building up soil fertility. First, it increases the soil fertility by the direct addition of nitrogen to the soil. Second, it improves the soil texture by the addition of humus or organic matter, which is essential for making the soil more productive. The addition of organic matter improves both heavy and sandy soils, as it has a binding effect on the loose particles of the sandy soils and makes the hard and heavy soils porous. Thus, it also increases the water-holding capacity of the soil. Besides, the conditions for increasing the number of useful bacteria in the soil are also improved.

Kind of Green Manuring

The practice of green manuring is performed in different ways according to suitable soil and climatic conditions of a particular area.

Broadly the practice of green manuring in India can be divided into two types:

1. *Green manuring in situ*: In this system, green manure crops are grown and buried in the same field, which is to be green manured, either as pure crop or an intercrop withy the main crop. The former system is followed in the northern India while latter is common in the central and eastern India.
2. *Green leaf manuring*: Green leaf manuring refers to turning into the soil green leave sand tender green twigs collected from shrubs and trees grown on bunds, wastelands and nearby forest areas. This system is generally followed in the central India.

The crop generally used for green manuring is dhaincha (Sesbania aculeata) though the cultivation of sun-hemp and guara is also in vogue. The crops commonly used for green-manuring in our country are the following:

Sunnhemp (crotolaraia juncea), Dhaincha (Sesbania aculeate), senji (Melilotus parviflora), Cowpea (Vigna catjang), berseem (Trifolium alexandrinum) etc. Sun hemp is the most outstanding green manure crop and is well suited in almost all parts of the country and fits in well with the sugarcane, potato, garden crops and the second season paddy in southern India and with irrigated wheat in the north. Dhaincha is also an outstanding green manure crop. It does well in the waterlogged and alkaline soil for it's reclamation programme. Green-manuring is in common use in irrigated lands, lands, but its popularity in barani land is hindered by the lack of irrigational facilities.

The Extension of Green-Manuring Scheme came into operation in the Punjab with effect from April 1, 1961. It aims at popularizing the use of green manure in the State. The Government encourages the adoption of this practice by the farmers by granting subsidies on seeds of green-manuring crops. The Irrigation Department also grants remission of water-rate, if crops are buried for green-manuring before 15th of September. The total area under the green-manuring crops in the district, during the past few years, has been as under:

Table. Area Under Green Manuring Crops with Time.

Year	Area Under Green-manuring Crops (Hectares)
1955-56	716
1960-61	7785
1965-66	13254
1967-68	26096

TECHNIQUES OF GREEN MANURING IN THE FIELD

The maximum benefit from the green manure crop cannot be obtained without knowing the:

- When it should be grown.
- When it should be buried into the soil.
- How much time should be given between the burying of the green manure crop and the sowing of the next crop.

TIME OF SOWING OF THE GREEN MANURE CROP

Normally the green manure crop should be grown immediately after the monsoon rains. As far as cultivation practice involved, no special care is needed in the preparation of the seedbed. Soil must have sufficient moisture for the quick germination and rapid early growth. Phosphatic fertilizers, if applied, should be evenly broadcast. Usually the seed of the green-manure crop is broadcast preferably with higher seed rate.

STAGE OF BURYING OF THE GREEN MANURE CROP

From the results of the several experiments, it is observed that best results of the green manuring are obtained if it is buried at the flowering stage. Majority of the crops take about 6 to 8 weeks to reach at the flowering stage from sowing.

Stage at which sun hemp was buried made a significant effect on the wheat yield. However basic principle is in green manuring crops, should aim at maximum succulent green matter at buring.

In one of the demonstration on effect of green manuring at the KAPURTHALA DISTRICT on paddy yield, we had observed that Greenmanuring increases the inherent soil fertility by fixing atmospheric nitrogen and by adding biomass there by saved 31. 0 to 36 per cent of urea as compared to control plots without having any adverse effect on the crop yield.

Time Interval between Burial of Green Manure Crop and the Sowing of the next Crop

The time interval should be allowed for complete decomposition of the turned in green manure crop before planning of the next crop and that time should depend upon the following factors:

1. Weather conditions
2. Nature of the buried green material in areas, receiving rainfall >50 inches humid conditions favors decomposition.

If the green manure crop is succulent, then there is no harm in transplanting the paddy immediately after turning in the green manure crop. However, in case of the woody, then sufficient time should be allowed for it's proper decomposition before planting the paddy. e. g. when succulent green manure crop of around 8 weeks was buried then paddy can be planted without having any adverse effect on the yield. But when dhaincha become woody (12 weeks), it was necessary to bury it about 4 to 8 weeks first for it's decomposition before planting paddy. In areas receiving 25 to 50 inches rainfall, green manure crop required about 6 to 8 weeks to decompose. It is only because of lesser moisture conditions. When green manure crop was intercropped in between the rows of the main crops like paddy, cotton, sugarcane etc. then it is buried in the succulent stage for it's rapid decomposition.

Plants Suitable for Green Manuring in the Field or in Situ

An ideal green manure crop should possess the following desired characteristics are as follows:

- It should be a legume with good nodular growth habit indicative of rapid nitrogen fixation under even unfavorable soil conditions.
- It should have little water requirements for it's own growth and should be capable of making a good stand on poor and exhausted soils.
- It should have a deep root system, which can be open the sub-soil and tap lower regions for plant nutrients.
- The plant should be of a leafy habit capable of producing heavy tender growth early in its life cycle.
- It should contain large quantities of non-fibrous tissues of rapid decomposability containing fair per cent of moisture and nitrogen.

Advantages of the Green manuring

Following are the some of the advantages of the green manuring:

1. It add organic matter to the soil. This stimulates the activity of the soil microorganisms.
2. The green manuring crops return to the upper soil plant nutrients taken up by the crop from deeper layers.

3. It improves the structure of the soil.
4. It facilitates the penetration of the rain water into the surface of the soil, thus decreasing the run-off and thus erosion.
5. The green manuring crops hold plant nutrients that would otherwise be lost by leaching.
6. When leguminous plants, like sun hemp and dhaincha are used as green manure crops, they add nitrogen to the soil for the succeeding crop
7. It increases the availability of certain plant nutrients like phosphorus (P2O5), calcium, potassium, magnesium and iron.

DISADVANTAGES OF THE GREEN MANURING

Every coin has got a head and a tail and it is the case with the green manuring. Some disadvantages are also associated with green manuring. When the proper technique of green manuring is not followed or when weather conditions become unfavorable, the following disadvantages are likely to become evident/happen.

1. Under rain fed conditions, it is feared that proper decomposition of the green manure crop and satisfactory germination of the succeeding crop may not take place if sufficient rainfall is not received after burying the green manure crop. This particularly applies to the wheat regions of the India.
2. Since green manuring for rabi season (Wheat) means the loss for the kharif crop, the practice of green manuring may not be always economical. This applies to the regions where irrigation facilities are available for raising kharif crop along with easy availability of fertilizers.
3. In case the main advantage of the green manuring is to be derived from addition of nitrogen, however sometimes the cost of growing green manure crops may be more than the cost of commercial nitrogen fertilizers.
4. An increase of diseases, insects and nematodes is possible.
5. A risk is involved in obtaining a satisfactory stand and growth of the green manure crops, if sufficient rainfall is not available.

Bibliography

A. C. Gaur : *Handbook of Organic Farming and Biofertilizers*, Ambica Book Agency, Jaipur, 2006.

A. C. Long : *Fish Feeding in Integrated Fish Farming*, Cyber Tech Publication, Delhi, 2012.

B. Ratna Kumari : *Farmers' Suicides in India: Impact on Women*, Serials Publication, Delhi, 2009.

B. S. Hansra and S.B. Shinde : *Farm Communication Through Mass Media in the New Millennium*, Agrotech Publication, Delhi, 2006.

D. Karpagam and S Ayisha Firdouse : *Dry Land Agriculture : Traditional Wisdom of Farmers for Sustainable Agriculture*, Agrobios Publication, Delhi, 2007.

D. M. Chandargi and K A Jahagirdar : *Extension Strategies for Promotion of Organic Farming*, Agrotech Publishing Academy, Udaipur, 2007.

D. P. Abrol and Uma Shankar : *Ecologically Based Integrated Pest Management*, New India Publishing Agency, Delhi, 2012.

D. P. Sarda : *Contract Farming and Farmers in India*, ABD Publication, Delhi, 2009.

D. V. Bhagat : *Encyclopaedia of Insect Pest Management*, Anmol Publication, Delhi, 2010.

F. R. Marshall : *Breeding Farm Animals*, Asiatic Publication, Delhi, 2006.

Frank, C. Edminster : *Fish Ponds for the Farm*, Agrobios Publication, Delhi, 2004.

G. D. Sharma : *Integrated Pest Management*, Swastik Publication, Delhi, 2010.

G. F. Warren : *Farm Management*, Arise Publication, Delhi, 2008.

. L. Sharma : *Phytonematode Management in Field Crops*, Oxford Book Company, 2009.

G. Raj Manohar and M. Murugan : *Emerging Opportunities in Alternative Poultry Farming Systems*, Satish Serial Publishing House, Delhi, 2013.

G. S. Dhaliwal and E.A. Heinrichs : *Critical Issues in Insect Pest Management*, Commonwealth, 1998.

G. Selvaraj and D. David Rajasekar : *Farm Diversification for Sustainable Agriculture*, International Book Distributor, 2002.

H. A. Modi : *Sustainable Organic Agriculture*, Aavishkar Publication, Delhi, 2012.

H. Panda : *Integrated Organic Farming Handbook*, Asia Pacific Business Press Inc., Delhi, 2013.

Hari Mohan Gupta : *Organic Farming and Sustainable Agriculture*, ABD Publication, Jaipur, 2005.

Harmeet Singh : *Dairy Farming*, APH Publication, Delhi, 2011.

J. P. J. Van Veren : *Soil Fertility and Sewage*, Agrobios Publication, Delhi, 2002.

K. Bhuvaneswari and S. Jaya Prabhavathi : *A Treatise on Integrated Pest Management*, Associated Publishing Company, 2012.

K.L. Blaxter and C.C. Thiel : *Farm Animals and their Management*, Satish Serial Publication, Delhi, 2005.

Laxmi Lal Somani : *Agri-Business and Farm Management*, Agrotech Publication, Delhi, 2010.

M. Israel Thomas and L. Nirmala : *Communication Techniques in Farm Extension*, Scientific Publication, Delhi, 2010.

N. C. Saxena and Vishwa Ballabh : *Farm Forestry in South Asia*, Sage Publication, Delhi, 2008.

N. K. Prasad : *Soil Fertility and Plant Nutrition*, IBDC Publishers, Delhi, 2013.

P. A. Koli and N.A. Patil : *Economics of Sericulture Farming*, University Book House, Delhi, 2004.

R. Quereshi : *Practical Organic Farming*, Arise Publication, Delhi, 2011.

R. Rajeswari and J. Prabhaharan : *A Handbook of Soil Fertility*, Satish Serial Publishing House, Delhi, 2013.

S. Arumugasamy and Subhashini Sridhar : *Community City Farming*, Earthcare Books, Delhi, 1998.

S. C. Agarwal : *A Hand Book of Fish Farming*, Narendra Publication, Delhi, 2007.

S. C. Panda : *Crop Management and Integrated Farming*, Agrobios Publication, Delhi, 2006.

T. K. Mohanty and S.S. Lathwal : *Data Mining Techniques for Farm Animal Management*, Agrotech Publishing Academy, Delhi, 2013.

T. Natarajan : *Organic Farming for Business*, Swastik Publications, New Delhi, 2011.

U. K. Behera : *Advances in Farming Systems*, Agrotech Publication, Delhi, 2012.

Vandana Shiva and Poonam Pandey : *Biodiversity based Organic Farming*, Earthcare Books, Kolkata, 2006.

Vandana Shiva and Poonam Pandey : *Biodiversity based Organic Farming*, Earthcare Books, Delhi, 2006.

Index